Sir John Lister-Kaye is one of Britain's best-known naturalists and conservationists. He is the author of eleven books on wildlife and the environment, including *The Dun Cow Rib*, shortlisted for the Wainwright Prize, and *Gods of the Morning*, winner of the Richard Jefferies Award for Nature Writing. John has lectured on the natural environment all over the world. He has served prominently in the RSPB, the Nature Conservancy Council, Scottish Natural Heritage and the Scottish Wildlife Trust. In 2003 he was awarded an OBE for services to nature conservation and in 2016 he was awarded the Royal Scottish Geographical Society's Geddes Environment Medal. He lives with his wife and family among the mountains of the Scottish Highlands, where he runs the world-famous Aigas Field Centre.

lister-kaye.co.uk | aigas.co.uk

Praise for John

'Utterly charming and captivating'
Sunday Times

'This book conjures otters, badgers, pine martens and weasels right onto the page, in language that is deft, vivid and alive'
JAY GRIFFITHS

'Does what all great nature writing should do: it makes you want to get out there yourself'
Mail on Sunday

'If only we could all be as attentive to the life around us as John Lister-Kaye. No one writes as movingly, or with such transporting poetic skills, about encounters with wild creatures'
HELEN MACDONALD

Also by John Lister-Kaye

Footprints in the Woods

The Secret Life of Forest and Riverbank

JOHN LISTER-KAYE

CANONGATE

This paperback edition published in 2024 by Canongate Books

First published in Great Britain, the USA and Canada in 2023
by Canongate Books Ltd,
14 High Street, Edinburgh EH1 1TE

Distributed in the USA by Publishers Group West
and in Canada by Publishers Group Canada

canongate.co.uk

1

British Library Cataloguing-in-Publication Data
A catalogue record for this book is available on
request from the British Library

ISBN 978 1 83885 880 3

Typeset in Dante MT by Palimpsest Book Production Ltd,
Falkirk, Stirlingshire

Printed and bound by CPI Group (UK) Ltd, Croydon CR0 4YY

For Harris and Gilby with love

My heart leaps up when I behold
A rainbow in the sky:
So was it when my life began;
So is it now I am a man;
So be it when I shall grow old,
Or let me die!

'My Heart Leaps Up',
William Wordsworth (1770–1850)

Contents

Foreword

'One touch of nature makes the whole world kin.'

Troilus and Cressida,
William Shakespeare (1564–1616)

Of all the dramatic wildlife we are fortunate enough to share our lives with – deer, wildcats, eagles, peregrine falcons, ospreys, red squirrels, many butterflies and wonderful plants – over and over again, for me, the very varied weasel family, *Mustelidae*, steals the show. Weasels and stoats, badgers and otters, and later on pine martens, were the stepping stones that would lead me to become a full-time naturalist and to want to work in nature conservation; and it was indirectly otters, arguably the most charismatic of our native Mustelids, that eventually led me to move to the Highlands and establish Scotland's first field studies centre at Aigas.

Badgers were the first real wild animals to grab my imagination as a boy reared in the pastoral countryside of 1950s England. There was something of the fabled 'wildwood' about their sandy mounds and deep, dark burrows. They spoke to me of ancient history, setts hundreds of generations old stretching back through the middle ages to 'days of old when knights were bold', a scion of Merrie England that had never altered, never given in to the breathless rush of human modernity. I loved them for that. Back then they seemed to me to possess an indefinable aura, drifting down through time like a long-abandoned way of life, and an irresistible wildness that drew me in and made me want to work with wildlife all my life.

Initially, they were all-consuming. Just as some people become obsessed with birds, so I became captivated by Mustelids even though I had no idea of their ecological significance, no real understanding of their fascinating natural history, far less their familial characteristics or taxonomy. Yet without realising it they were teaching me to become feral myself, guiding me into their ways so that my senses seemed to merge with theirs at a level far more intimate than just scientific interest or admiration.

I

Place and Time

'Those who contemplate the beauty of the earth find reserves of strength that will endure as long as life lasts. There is something infinitely healing in the repeated refrains of nature – the assurance that dawn comes after night, and spring after winter.'

The Sense of Wonder,
Rachel Carson (1907–1964)

Oh! for a landscape of dreams and a dream of wildness uncontained. Long ago this place gripped me and has never let go. Here at Aigas we are ringed in by mountains, moors and forests. The mountains are constants, unchanging, dependable and bountiful. They bring assurance and conviction to our days. They can claim your spirit and lock it away for safekeeping, whereas the moors and forests drift with the seasons and have stories of their own. They award us texture and an endless, shifting beauty. Even after all these years, every day offers up a familiar and yet exciting freshness. Every morning I awaken in expectation. What bright and startling gems will each new dawn reveal?

All night the window stays wide open beside our bed – storm, sleet, blizzard, driving rain or moonshine. It's a habit that mildly unsettled my wife, Lucy, when we married more than thirty years ago. Now she just accepts that part of me never completely comes indoors, that, like the roe deer lying up in the broom thicket on the edge of the moor, I need to remain tuned in to whatever elemental tidings the winds disclose.

My first act of morning is to stand at that window, to feel the breeze on my face and stare, my mind spinning and drifting like a sycamore seed, hoping for a revelation. Mostly, of course, there is nothing unusual. Yet just the passive act of listening and looking is a tacit acknowledgement that we live in two entirely separate worlds – worlds we can never hope to reconcile. Ours is the world of wonder held at arm's length from the vital world outside. Awakening to a new day and shrugging off the mantle of sleep is in itself an act of renewal, as much of hope as of promise.

Sometimes there is a bird alarming, such as the edged, chip-chipping of a blackbird, or perhaps an empty silence suggesting that a sparrowhawk has just passed through, streaming a trail of fear through the gardens. I feel my awareness sharpen, unlocking the moment. If I'm lucky and catch sight of that sparrowhawk I am instantly awake, all senses flaring. Sharing the moment is thrilling, with an electric vitality all its own. I long to be out there, living it.

★   ★   ★

'Nature without check with original energy.'

'Song of Myself',
Walt Whitman (1819–1892)

Wildlife needs habitat. For survival the two are inseparable. Habitat is a portmanteau term; it covers all bases. All wild creatures require a natural home, a place and an environment where they can feed, breed and flourish.

A habitat can be wet or dry, hot or cold, light or dark, at the bottom of the ocean or on the summit of high mountains and everything in between. A mayfly needs clean, fresh water in which to lay its eggs and where its larvae can thrive. Butterflies and moths need the right food plants for their caterpillars. Red squirrels need forests and woods for sanctuary where they can build their leaf and twig-bundle dreys, the trees delivering the right seeds and nuts. Plants need particular soil and light conditions and access to moisture and nutrients. Otters need fish-full rivers, lakes and foreshore. Habitat is everything.

Perhaps the best definition of habitat is just that: a natural home. If humans destroy habitat, wildlife either has to move somewhere else, adapt to new conditions or perish. If the butterfly's food plant is removed, so is the butterfly. We are exceptionally good at destroying habitat for wildlife; we excel at it – it is what human beings have always done ever since we stepped out of the primeval forest and parted company with the wild. Sadly, far too often creating habitat for people means destroying it for just about everything else. The principal cause cited by scientists for wildlife decline is habitat loss, brought about by our endlessly expanding needs and activities. The mantra of wildlife destruction has been repeated so often it has become stale: industrial agriculture, commercial forestry, urbanisation, pollution, over-fishing . . . the list goes on and on, all now hugely complicated by man-accelerated climate change. Wildlife is always a secondary consideration – or often no consideration at all.

Yet there are still a few places where the topography and the nature of the land don't make convenient habitat for humans: places such as mountains, coastal cliffs, windswept moors, steep forests, bogs, marshes and riverbanks and floodplains. On those sites wildlife can find a home, a natural home, and can feed, breed and thrive. Despite man's best efforts to tame it, the rugged uplands of the Highlands of Scotland have always provided habitat for wildlife. Fifty years ago that seemed to me to be a good reason for choosing to live here and to create a field studies centre. To have lived those fifty years surrounded by wildlife, and in many instances to have shared their habitat with them, has been an existential privilege and a joy. It has also been my life's work.

*   *   *

The house stands solid and slightly smug, but also randomly perched, as you might hurriedly plonk a pile of books on the edge of a table. Smug because, despite its elaborate appearance, it has stood here for many centuries and wears that knowing expression of long ownership, weighty with history; and perched because the land falls away from it in irregular, rippling terraces, some steep, some slight, down to the winding river three hundred feet below.

We sit on the partially collapsed wall of a glacial glen – that word 'glen' burdened with the turbulent and emotive history of a native people long gone. Below us it trails twenty-two miles southwest, back up to the nineteen 'Munro' peaks of the Affric Mountains, from whence the glaciers emerged. It is a deep ice-carved trough snaking through steep rock walls thinly clothed in brown mineral soils where broom and gorse flowers, egg-yolk yellow, lace the spring sunshine with heady bouts of coconut. The glen is also where people live.

Just here, where our windows gaze out over the valley, massive forces shaped our world. Twenty-five thousand years ago the advancing glacier hit a fault line between hard, grey metamorphic schist and a much more friable conglomerate rock composed of water-smoothed pebbles from some long-lost shore-line, now welded into heat-seared sandstone. For many millennia the huge ice body, 3,000 feet thick, half a mile wide and twenty miles long, creaked and groaned as its colossal bulk turned half right to follow the fault – the line of least resistance – grinding and shattering the conglomerate wall as it gouged its way past and on out to what is now the Moray Firth and the North Sea.

As the ice began to retreat around 12,000 years ago, to melt back up its U-shaped valley, its surging summer meltwater generously handed us a beneficent legacy. Those rushing streams washed out the shattered valley wall along the fault line, right here, where the house stands, depositing its spoils in smooth gravel and boulder-clay moraine terraces, now handsomely robed as precious woods and fields.

Once the glacier had melted away, the river claimed the valley. It now commands our view. A long time ago someone gave it a name – the River Beauly, fed by the confluence of the rivers Glass and Farrar three miles upstream. Its dark waters have taken the glen over, languidly winding through the many meanders of its flood plain of damp fields and marshes with the autocratic air of long-held authority. Ox-bow lakes reveal the forgotten sagas of its many changing moods. In the 1950s the Beauly was dammed for hydro power, slowing its flow; now gnarled alders line the banks, twisted and bowed like old men, limbs stretching low over dark water.

When heavy spring rains combine with melting snows in the mountains to hurl the river's many tributaries and steep feeder burns into crashing, roaring spate, the valley fields flood. I have seen the whole glen awash from glacial wall to glacial wall, crofters' cattle and sheep marooned among the alders on slender islets of raised moraine, creamy-white water surging by. I once witnessed a bitch otter taking her cubs to safety in a roadside culvert high above the torrent, one cub hanging from her jaws, two more stumbling along behind.

After the floods recede, unless the pastures are soon tended, they run quickly to dense rushes, whose waterborne seeds chart the flood margins in swathes of moss green, dark and revealing. They expose the industrious and the indolent. To drive the glen road a few months later is to witness nature's raw appraisal – we can see who nurtures their river fields and who does not.

The road hugs the valley edge past lonely croft cottages. It serves the tiny estate settlement of Struy, for centuries the tribal heartland of Clan Chisholm, then a few miles further on to the tight little crossroads village of Cannich. With the canny insight of generations of Highland folk, roads and crofts were both sited just above the flood line. There is upland farming and crofting on the soggy river meadows, mostly cattle and sheep grazings and occasional crops of hay or silage, sometimes turnips as winter feed for sheep – and always uneconomic. This land does not yield willingly to human needs.

Above the steep glen walls another world awaits. Largely unseen from the road, on an undulating blanket of waterlogged peat the stretching moors rise hazily to a beckoning, far-off horizon of much higher hills and mountains, like a distant land of make-believe and childhood imagination. On grey days we feel hemmed in. The hills lift to bruising ramparts of cloud so that you can no longer tell where land ends and sky takes over.

If, on brighter days, I climb to the trig point on the little hill above the moor, I can see the massive, nearly 4,000-foot hulk of Sgùrr na Lapaich, 'rocky peak of the bog' in the native Gaelic, seventeen miles to the southwest. Its deep snows blush rose-pink in low winter sun; in its shady north-facing corries snowdrifts can linger patchily well into the warmth of June, sometimes dotted with red deer hinds cooling off. In high summer the mountain is mantled in the grey-greens of lichen, or anywhere from the softest dove grey through a blurry plum bloom to scowling charcoal as the light clocks round its many moods and faces, never the same two days in a row.

Highland light has a character all its own. Latitude dictates our dutiful track around the sun, never overhead this far north. It awards us our seasons. Low-angled light and the absence of dust in the atmosphere allow an intensity and fierce clarity unknown further south. Dawns gather behind the hills like raiding clans, finally spilling in and overwhelming the glen with verve and flair. Even on cloudy days the sun fools with us, suddenly bursting through, spotlighting a startling patch of hill or painting the loch with brilliant stripes of silver. Sometimes it trails across the river fields, teasing the eye to follow it, much as a nesting lapwing feigns a broken wing.

For most of the year coarse grasses pale the moors to chartreuse – in some lights yellow, sometimes lettuce-green – and the burnt umber of ling heather scowls from the dryer knolls and ridges. When rain clouds occlude the sun, their sinister shadows surf the moors like hostile battleships. For many centuries these windswept uplands were the essential summer shielings for Highlanders' cattle and sheep, but the days of transhumance are long gone, as are most of the people, who abandoned the glen in the nineteenth century for the promise of a better life in the New World. Relieved of grazing pressure, the moors have gone wild again.

Delicate acid-loving flowers abound: the sky-blue of milkwort and tiny yellow tormentil flowers grace the spring moors with the sticky-leaved insect-absorbing sundews and butterwort. Tufts of fluffy bog cotton glow like stars in wet flushes, the crinkly leaves of lousewort's sober pink, and sheets of golden bog asphodel mix with northern marsh and heath-spotted orchids among coarse moorland grasses. In a few secret places in the pinewoods the exquisitely delicate pink-white bells of twinflower steal the show.

When, in August, the heather blooms, an episcopal haze settles around us. But there is competition for this land, a struggle never won. On the better-drained soils, as upland sheep became uneconomic and grazing receded, bracken has invaded with all the impudence of a spiteful neighbour, leaving little hope for anything else above or below ground. Like tangled black boot-laces, its menacing rhizomes weave an impenetrable mesh in the peaty soil, strangling all comers at birth. Only the occasional Scots pine, rowan tree or juniper bush have somehow grabbed a niche, prised open a foothold, and now also dot the view. Those trees award depth and perspective to an otherwise forlorn moorland landscape.

There is nothing forlorn about the woods edging the moors: ash, sessile oak, rowan, Scots pine, goat and eared willows, holly, hazel and gean (wild cherry) all prosper here, but it is the downy birch, *Betula pubescens*, northern cousin to the silver birch, which flourishes thick and firm-limbed on acid soils hard won from the heavy underlying boulder clays, ice-layered by the glacial outwash. Our birchwoods are a blessing; they shade out the bracken and grace the hill slopes with gentle quilts of fertility, their welcome calories building the richest soils in leaf litter, allowing humus and secret networks of mycorrhizal fungi to accrete, while sheets of wood anemone, wood sorrel and bluebells float sensuously in their dappling shade. Every spring a chorus of birdsong rings out from our woods.

Wind-borne birch seeds, wafted in on prevailing south-westerlies, are always quick to take over any broken or exposed ground. They germinate in a flash, rushing to the sunlight in thickets of brutal competition. The winners grow tall and straight, gleaming white-barked until maturity, but in old age the lower trunks erupt and split into dark crusts of rough bark fissures much loved by spiders, coal tits and treecreepers.

Leaf-buds and twigs of downy birch are of burnt sienna. Viewed across the winter valley those leafless woods glow a luminous chocolate-purple in the low-angled sun. Birches are not long lived; in senescence most succumb to polypore bracket fungi or to wind-throw before they reach a hundred. Draped with green-grey lichens their rotting trunks are sanctuary to the eggs and larvae of many bugs and can be readily hollowed by nesting greater-spotted woodpeckers. Birch is a generous hardwood, honest and true; it seems to grace its omni-presence in our woods by giving back. Stacked to dry, it delivers splendid, clean-splitting firewood and the after-glow of long winter evenings beside the fire.

In a few scattered places there are small Scots pine-woods, the once-dominant tree of Highland forests, much exploited for their resinous, slow-grown heart-wood, as strong as steel. These relict copses, some of only a few isolated acres, are a vital link with the far larger, ancient Caledonian pinewoods in glens Affric,

Cannich and Strathfarrar to the south and west. They are staging posts for specialist species such as crested tits and crossbills, as well as red squirrels and an unseen host of invertebrates.

Those old pinewoods are abbeys among chapels of lesser woodland. To stand in their midst is to wilt into timelessness. They clutch at the soul, stirring brave emotions and banishing misgivings. Individual 'granny' pines, often contorted with age, seem to defy their own genes, twisting and lurching more akin to venerable olive trees than pines. They exude aura, compelling you to look again. You soften at their feet.

One such copse of only five or six acres, self-regenerated after wartime clearance, skirts the shore of the little loch above the house. Summer or winter it never fails to delight. It is where woodcock probe the needly floor for worms and every year sparrowhawks return to repair the old crow's nest where they have raised their squeaky broods for more years than I can remember. It is also where every May roe does drop their twin fawns and where, just once, a wildcat stalked silently past me in the dawn.

It is a lifetime ago, more than fifty years, that chance brought me to live in the Highlands, although it doesn't seem that long. Half a century of watching the seasons roll over me like tides, of yielding to the compulsive wildness of the place. It is a landscape that has taken me over, fashioned me to its ways. I now view it – no, *feel* it – as a land apart from the rest of Britain, a place of richly intertwined habitats where nature still manages to cock a snoot at mankind's ceaseless and often mindless efforts to tame it and bring it to heel.

Nature is persistent and has plans of its own. It never sleeps. Like the rushes on the river meadows, broom and gorse defiantly shoot up and will quickly overwhelm any disturbed land. Every spring they crowd the valley walls with bright yellow blooms – the same startling yellow as wild flags and globe flowers – blooms so densely packed that the green vanishes altogether; in high summer they crackle with bursting seedpods, firing little black peas in every direction. Turn your back and they've germinated, rising phoenix-like from the winter-killed grass in a spiky green fog. Left alone for a few years they conspire to create a jungle thicket eight feet high and so dense that everything else is smothered – a barricade of abandonment and a colourful nod to man's surrender.

This is a land of shafting light and the sound of falling water; in summer a gentle burble, in winter the roar of outrage uncontained. We live in thrall to the seasons, a bondage we would not be without. Spring hovers at the door, reluctant to enter. When it makes up its mind, it's a festival of birdsong, a surge of elation and happy memories reawakened. It burns bright after the long dark and flares like a Roman candle, fizzing out before you've really taken it in. The rush of verdure spurred by ever-lengthening daylight flings us headlong into the helter-skelter of summer. It races to the solstice.

Twilight stretches and yawns like a sleepy drunk. Nights dwindle to a brief sigh, nocturnes of little more than an hour, perfect for poachers. Dawns creep back in, sneaky as a windless tide.

Summer livery is all green; by mid-July a verdant drabness pervades the land like a dire edict. August days drag themselves from week to week, lethargic and laden. Lengthening nights are burdened with restless cumulonimbus; unspeakably troubled thunderheads rumble and churn high above us until lances of fork lightning cleave the static darkness. Warm rain deluges through in sudden pulses and is gone again, leaving us gulping fresher air. The wildflowers melt away. But summer clutches at September in vain. The challenges of belling red deer stags echo from the hills before the month is out, and with the first moonlit frosts rust canters through the bracken.

Dawn mists flood the whole valley into a linear silver lake, gleaming in first sun. The deer grass in the hills bleeds its green away to a uniform biscuit-ginger and urgency grips the dragonflies at the loch. They hurry to dip their eggs between browning water-lily pads about to sink. For a few weeks the birchwoods shimmer gold in a cool and lowering sun and bright clusters of scarlet rowanberries bring a lipstick glamour to the woods. The chlorophyll verdure of leaves has bled silently away, revealing for a week or two before they fall the startling gold and firebrand oranges of autumn hidden below. When the first over-wintering geese straggle across burdened clouds, their wild music winnowing down from a turbulent sky, we know that winter is on its way.

Mellow autumn slips quietly away and darkness falls with the last umber leaves drifting to earth. Winter commands six months, some years clutching at seven. It shuts us down. Days are so short that if you think twice about an outdoor job, you've missed your chance. If it shows its face at all, an embarrassed sun arrives late, skirts low, not-ought-to-be-here-like along the hill-tops and quickly ducks out of sight, leaving us with the chill white sky of the north. Night descends, burgling the daylight away like a greedy thief.

On clear nights the stars drill down, the moon lifts and glides through air as pure as gin, skimming the black water of the loch with a mercurial sheen, and, when you least expect them, the northern lights, the *aurora borealis*, flare and wave in luminous arcs of lime and burgundy, spinning you in reverential circles. Their silence drums from horizon to horizon.

Winter. The pulses of life have all slowed. Only the robins pierce the calm, not the ringing joys of spring, but a strained, repeating tinkle like an augury, the dire portent of still darker days to come. Snows flurry in on bitter winds, shrouding the high hills and revealing the lacework of fox tracks across the fields. The loch freezes over, some years thick enough to skate, and the ice booms and crackles like far-off cannon fire.

Red squirrels wrap themselves tighter in their fluffy tails, tucked away in dreys high up in the pinewood. The snoring badger is curled in a bed of grasses in the depths of his underground fastness. Rime dulls the rose hips' shine and leaves are crisp beneath our feet. Under the roof slates the pipistrelle bats are silently sleeping, dreaming of a mild spell when they can cast off their torpor and nip out for a quick moth. In the hen house the cockerels crow out from the darkness, lamenting dawns that used to be so prompt.

Such are the dimensions of the land that surrounds us and shapes our lives. The long winding river, the deep glen and its wooded slopes, and the wide, empty moors all contain us and wrap us round. They banish our fictions and let them burn away in sunlight. They hold us in our place. It is a land that makes us think with our hands; it dictates our apparel and fires our language, our moods, our hopes and our dreams. It takes charge. But much more than that, with the fourth dimension, time, it provides an uncountable multitude of habitats for wildlife. It awards us neighbours.

Humans cannot dictate to the clouds that gather round the snowy summit of the mountain, that feed the river and its tributary burns rushing through the glen, although we like to pretend we're in charge. Like the orbit of the moon and the boundless dome of the stars, they are truly wild and way beyond our imperious regulation. The river and the moors and mountains are as close to the notion of Wild Nature as we can get in Britain, and with that precious wildness comes life – wildlife – the self-willed creatures with whom we share the land and our days. This is a world on the edge and it begs for the outdoor life. For man and beast alike it is one heck of a place to live.

There are deer, of course, herds of wild red deer in the hills, a feature of the Highlands everyone expects, and the delicate roe whose plucking and browsing incisors subtly sculpt the character of the woods. Otters stealthily patrol the river and launch nocturnal raids on the loch's brown trout. Red squirrels chase each other round scaly pine trunks and scatter cones across the needly floor. The sand-yellow mounds of ancient badger setts dot the woodland glades and rarely – very rarely – we hear of wildcats denning among the rocks of the steep glen wall. And then there are the birds – thank heaven for the birds. Even in the depths of winter they serenade our dawns and spill life and movement into all our days.

Golden eagles in the mountains; ospreys crashing into lochs and rivers; the screams of peregrines echoing from the river gorge; the guttural cronk of ravens; the chip-chat of crossbills trooping over the pinewood crowns; the loop and flicker of woodcock wings between the birches; the wobbling, sneezing taunts of blackcock lekking on the moorland edges; barn owls silently floating over the river meadows at dusk; the dashing sparrowhawk and the darting merlin; the heron's rough crake; the bugling sonnets of whooper swans arriving for winter and the seasonal haggle of pink-footed geese in broad chevrons levering through our spring and autumn skies . . . These are our neighbours, the visible wildlife – the mammals and birds that gladden and nourish our Highland days. And yet there are others not so readily encountered. There are tell-tale footprints in the woods, a fleeting blur of chestnut fur, rootlings in the morning lawns, eyes momentarily flaring in a torch beam, a smoke curl of tail vanishing behind a building and screams of outrage piercing the moonlit night. We are never alone.

2

The Weasel Family

Mustelidae: weasels, stoats, mink, martens, polecats, otters, badgers, wolverines – altogether some sixty-five species across the globe – is the largest family in the order *Carnivora*. Mostly they are flesh-eating although some, like the badgers and many of the martens, are very omnivorous, eaters of as much fruit, nuts and other vegetable matter as of animal prey.

Here at Aigas we have four and occasionally five of the above: weasels, pine martens, otters and badgers have strong populations across the Highlands, with stoats, *Mustela erminea*, only rarely putting in an appearance locally (they used to be regular, but have disappeared with their principal prey, rabbits, eradicated by imported disease). Polecats, *Mustela putorius*, were hunted out for their mink-like fur in the eighteenth and nineteenth centuries but have clung on in Wales and now are enjoying a gradual return to the Scottish Borders.

Mink, *Neovison vison*, are not native to Britain. They are exotic escapees from fur farms and originated in North America. Fur farming is now illegal in the UK. Although they are widespread, they are considered a pest, seriously threatening our native water voles, raiding fish stocks and damaging wider biodiversity. They are semi-aquatic, extremely versatile predators, as much at home in the sea as in rivers and streams. But it is the weasel that has given his name to the whole family.

> Thin as death, the dark
> brown weasel slides
> like smoke through night's hard silence.
> The worlds of the small are still. He glides
> beneath the chicken house. Bird life
> above him sleeps in feathers as he creeps
> among the stones, small nose testing every board
> for opening, a hole as small as an eye, a fallen knot,
> a crack where time has broken through.
>
> from 'Weasel',
> Patrick Lane (1939–2019)

Mustela nivalis, the snowy weasel. Latin *mustela*, a weasel; the specific adjective *nivalis*, snowy. Old English *wesle, wesule*; Dutch *wezel*; German *wiesel*. The 'snowy' weasel because in some parts of its northern hemisphere range weasels turn pure white, their aggressive camouflage of winter. In Britain ours don't, but they do moult from a summer chestnut above and off-white below into a thicker winter pelage of brighter white underside, the upper body emerging a glowing fruitwood brown.

Weasels are tiny. Think mouse, elongated to 20–23cm with short legs and 6–7cm of short tail, and with fangs – not to be confused with the stoat, which, although very similar in shape, is more than twice the size with a much longer, black-tipped tail. Female weasels are smaller than males, but identical in appearance. Their fur is a fine bright chestnut above and clear white below from under the chin almost down to the tail, although the demarcation line between back and belly colour varies greatly between individuals. They moult twice a year, spring and autumn, and albinos are not uncommon.

In spring or summer they give birth to one litter of up to six kits and the bitch raises the young, teaching them to hunt. In common with other mustelids, weasels possess a scent gland beneath the tail, used for identification and marking territory. It gives them a strong, clearly identifiable musky smell. Our weasel is found throughout Europe and as far as Siberia, China and Afghanistan. It occurs widely in England, Scotland and Wales, but is absent from Ireland, the Scilly Isles, the Hebrides, Orkney and Shetland.

Typical habitat is woodland, hedgerows, scrubland, along streams and rivers, in suburban gardens, on open heath and moorland, in fact anywhere containing prey and cover. An apparently liquid body that can squeeze into mouse tunnels must be a mouse's worst nightmare. A weasel skull can pass through a large signet ring and the rest of the sinuous body can follow. But no mouse likes to think weasel. Pound for pound the weasel ranks with the fiercest carnivores on earth. They are killers, leopards of the underworld, hunting down mice, rats and voles, but also fearlessly tackling prey far larger than themselves: rabbits and leverets up to twenty times their weight fall to their ferocity – few other carnivores do that – and birds as big as pheasants. The full prey list is extensive: lizards, toads, frogs, slow worms, the eggs of ground-nesting birds, shrews, moles, squirrels, crayfish and insects, but probably 90 per cent is mice and voles, making weasel populations expand and contract with those of the principal prey species. In hard times weasels have been known to take carrion.

They are not long-lived; the list of potential predators is long and relentless. Although in captivity they have been known to survive for ten years, in the hazardous wild most will live for only three to five years. They can reproduce in their first year, gestation is four weeks and the female will raise her kits in a nest of grasses, feathers and mouse fur, often in a den in the roots of a tree, or inside a dry-stone wall, a hollow log or stacks of cut firewood, almost anywhere dry and secure.

★　　★　　★

You cannot stalk a weasel. It is too sharp, too alert, too bright-eyed, too fired and sprung. It is also too persecuted. The shadow of man is deeply etched into that fizzing little brain. A weasel can vanish in a second. You have to wait for it to come to you – and that may never happen. I had one once, a foundling weasel, raised from a tiny kit to a fully charged adult male, reluctantly released back into the wild at a year old. I learned to love weasels and that weasel taught me weasel ways I have never forgotten. Setting him free was one of the hardest things I have ever had to do. The passing years have neither dimmed nor leached those memories away.

18 March: midday

Lying snow, wet with petulant skelps of cheek-stinging sleet and rain. Slush under skies of a turbulent, oceanic grey. A land suspended: too mild to freeze, too cold to thaw. A funereal gloom gathered and held. In the wind the birchwoods dripped and I shivered. Nothing moved except the tops of trees veering from a sly north-easterly, sabre-sharp and pitiless. Not even the hoodie crows, ubiquitous villains of the Highlands, emerged to challenge the day. They live on the edge, those hoodies, lives of calculated Machiavellian brinkmanship. Often a pair follows me, warily, at a safe distance – they know very well the range of a gun – calling to each other in a rasping monotony of self-assuring blasphemy. Highly intelligent and always opportunistically and malevolently curious, not just to know what I am doing, but to see what else I might disturb – what pickings chance might deliver up. Not today. The rank grasses at the edge of the wood swished wetly against my boots as if wading through a marsh. I ducked into the trees for the shreds of shelter they could offer.

At the old dry-stone dyke, the boundary between what had once been croft grazings, I clambered over in a gap of tumbled copestones and sat huddled in the welcome half-dry of its lee. Its friendly boulders seemed to caress my leaning back. Sleet continued to bluster past high above my head, rattling the topmost bare twigs of the birches. At ground level a benevolent calm settled at my feet.

The wall, mossy and with tufts of hard fern sprouting from between its basal boulders, curved away uphill to my right. I thought of the thousands of hours of back-breaking labour invested in that old structure; the lugging and levering into place, the rough hands, winter breath pluming from growled curses and oaths, the sledge-heaving ponies sweating at the neck, hocks straining, shouts ringing through the wood. Was it a hundred years, I wondered, two hundred or much, much older? Long enough for drumlins of dark moss to wrap and rise on the top stones, and for discs of beady ginger lichens to scroll timelessly across surfaces the sun never saw. Such prehistoric organisms have no calendar; they disregard time. To them our lives are a flicker of falling leaves.

A movement. Eyes reined in tight, fixed on a darkness between boulders ten feet away at the base of the wall. Was that a mouse, or perhaps a vole? Or did I just imagine it? But it wasn't either of those, and deep inside I knew it. I knew it because instinct and the long memory spoke weasel to me loud and clear – and I knew it would return. The angular little face, hazelnut-brown; eyes as bright as beads. Flash of white from the throat down to the belly. I froze.

*   *   *

That long-ago weasel nest had been in a fallen elm log, hollow, ivy-covered and many decades dead. It proved a bad choice. A badger's hypersensitive nose had sniffed it out and torn the ivy and rotten wood away with ripping canines and long front claws, mechanically strong. The mother weasel had fled. Even the valiant heart of a weasel bitch is no match for a hungry badger. It cleaned up – the nest rent asunder, three, four, maybe five kits? One missed. Tiny, eyes barely open, less than three weeks old, unable to walk or crawl, the oh-so-helpless scrap of weasel was cast aside in a shredded bundle of moss and grass.

The drama was plain enough, only a few yards from a woodland path. The unmistakable havoc of a badger's rapacious plunder spread wide. Rotten timber mouthed and flung, raked moss among spears of trampled ivy and wood-rush, the log clawed deep and bare. It drew me in. Then I heard it. The feeblest, tissue-thin animal cry you can imagine, a demi-bleat of sound from somewhere beneath my feet. The moss shreds parted. Two inches long, a slender twist of chestnut fur in perfect weasel shape, blind, blunt-nosed and doomed. I was a teenage schoolboy, entranced and immediately determined to rescue this scrap of a life barely begun.

★   ★   ★

That was then. Now, half a century later, I sat staring at this old dyke. Walls are irresistible to weasels as long as they are dry-stone. Highland dykes are rarely more than two large boulders thick with an in-fill of rubble between, narrowing as they rise to a single cope across the top. Broad linkstones bridge and stabilise the middle courses, all packed round with more small-stone rubble and whatever gravelly debris is to hand. Such walls are skillfully built with the wisdom of age-old hands and eyes, talents passed down. They are defiantly stable and will stand square-shouldered against the elements for centuries. As the decades pass, frost loosens the structural boulders and rainfall and melting snows wash through, flushing out and sifting down the loose in-fill. Cavities emerge, unplanned channels, corridors, holes and galleries, perfect for mice – and for weasels.

Repairing a fallen length of dyke a few years ago, we had to dismantle it first. Deep in its interior I found a weasel nest. My helper thought it was a mouse nest, but when I bent to inspect it and parted the moss and grass bundle, the heavy musteline reek spooled me back to those hallowed days of boyhood when my weasel lived in the pocket of my tweed jacket. Unforgettable, unmistakable, irresistible; all the power of recall the vivid tincture of a scent can muster. I closed my eyes and breathed deeply. If that fragrance could be a fabric it would be a deep, glowing purple or magenta velvet, the arras drapes of some ancient basilica. Worn as a robe it would speak of power and wealth. Small wonder the perfume industry was built on animal musk.

Inside that dyke we had also found the shards of many mouse and vole kills: bones and fur, tiny feet and some skulls and teeth scattered around, like totems in an ancient tomb. Because the interiors of such walls are sheltered and dry, mice, particularly wood mice, *Apodemus*, nest there too. It doesn't say much for mouse intelligence – sheep choosing to shelter in the she-wolf's den. You would think, wouldn't you, that after all the multi-millennia of weasel predation that mice would know, or learn, or remember, or recognise that indelible scent, that glaring, unequivocal, trumpeting incense of certain death? A signal that shouts 'Go!' But no, mice continue to frequent such walls and die igno-miniously in the skull-crunching jaws of their most feared and dreaded predator – the weasel.

So I sat, and stared, and waited, bright memories of the distant past honing the moment, all senses charged. Most species of *Mustelidae* are inquisitive by nature, always on the *qui vive*. Driven by genes primed with a restless, opportunistic spirit, they quiver with a need to know what's going on. Sit still and they are certain to check you out. I knew it wouldn't take long. Nature solicits serendipity. Knowledge and experience glide into instinct: weasels love old walls; they are irresistibly nosy and must believe those internal caverns and galleries are made just for them. Charged with suspicion and a primed, wary intelligence, they seldom reappear in the same place twice. Sooner or later this bright little weasel would check me out. Three minutes or four? Maybe less. Time drags when expectation plucks at your patience. Only my eyes moved, scanning gently, steadily, least movement possible, up and across that dozen yards of wall.

A bit closer, I had thought – maybe the distance halved. Wrong. Suddenly there, right there, not ten feet away, and not just a face. Two shining eyes and little round bilberry-leaf ears rimmed in white, the tiny nose as pinky-brown as a peanut, a gleaming white bib between front paws determinedly placed. Staring straight at me.

In a sprawling willow a wren alarmed loud and shrill, a prattling, rattling trill, insistent and repeating over and over like a mechanical toy. It had also spotted the movement and instantly knew fear. It flitted nervously from branch to branch. The weasel ignored it. Weasels know wrens and wrens know weasels. Wrens harass weasels all the time, ever fearful that the predator will find their nest, perhaps also tucked into a wall cavity. But this weasel was not interested in wrens. He was checking out a much more threatening presence. Man the dread predator. Man the dark shadow to all wildlife. Man who preys on predators with traps and snares and guns, who lays poison, purges the land with vile chemicals and layers the world with concrete and tarmac. Man the destroyer of habitats. We are the killers and every creature out there knows it. Our murderous reputation is older than the birch trees, older than the wall, older than the oldest pine, older even than the landscape itself. It pre-dates the glaciers and the winding river, the dark mosses and the ginger lichens so unfathomably slow to grow. We stink of death wherever we go. There can be no denial.

But was I human? Was this huddled shape, silent and unmoving, the dread predator? No wind in that sheltered place, no swirling death scent; no hint of trap or gun; no familiar outline to the crouched form. A puzzle. Even at ten feet a conundrum was flaring in that fierce little brain, a burning weasel dilemma. Then gone. Vanished again, back into the wall as if it had never been there. I sat still, blanked and wondering. But I had not detected fear. I had seen no panic in those little jet eyes, no sense of alarm or recognition or the puzzle resolved. I sat still. This weasel was still checking me out.

Twenty seconds later it was back, now slightly further out, peering again and at my eye level. The short, muscled neck leading into the flowing back, the white rippling underside edges, the full stretch of bib. Fixed. It felt like a game – who will flinch first. A minute drifted by. We stared at each other locked, or dazed, or stunned into an impasse, an awkward, unspoken truce. I wanted to offer a deal – if I promise not to be human, can we be quits? Could you conquer your dread, little weasel? Will you let me into your weasel world? Is there anything I can do to throw off the mask of terror? But no, that was too much even to think. It was as though it could read my mind. Contempt for such folly broke like a wave inside its tight little skull and in the blink of an eye it was gone, turned and vanished back into the wall.

Then it dawned. Perhaps I was in the way. Was I precisely where that weasel needed to be? Was there a weasel nest inside the wall behind me? Had I plonked myself down to shelter only inches from its carefully chosen den? Was she a bitch weasel with young, newborn, blind and helpless, swaddled in moss, mouse fur and feathers somewhere in the dry wall cavities behind me? Is that what was driving its boldness, its apparent fearlessness? I moved quickly and silently to a birch tree a few yards away and sat down again, now facing the wall.

★ ★ ★

My weasel grew strong on diluted cows' milk and glucose through a pipette, which he grasped eagerly with his fore paws, lying on his back. I fed him every few hours night and day; kept him in a small box no bigger than a fat book, a cotton wool nest inside. From the very beginning he was meticulously clean. He would clamber clumsily to the edge of the box and, with a look of intense concentration on his tiny blunt face, he would produce a minuscule caterpillar of oily black excrement, always in the same place, always outside the nest. I named him Wilba and he slept in my bed, the box close to my pillow so that I could wake and attend to his thin cries when he was hungry. He grew quickly, becoming responsive and playful, the envy of many other boys in my school. In only a few weeks he was a proper weasel: alert, bright-eyed, his mustelid intelligence shining through a shared adolescence, his as rapid as a sunrise, mine dawdling along, entranced, happy to live out each stretched day of learning.

When he was awake he demanded constant attention. Although never a pet, it was unavoidable that he was utterly dependent upon me and I carried him with me everywhere I went. He would sleep happily in the pocket of my tweed jacket – the required uniform for most 1950s and '60s boarding schools – and as he grew I adjusted the lining of the jacket to create a cloth tunnel from my left-hand pocket up inside the jacket to the breast pocket. When he awoke I would feel him tunnelling up until suddenly his bright-eyed little face would appear, happily chittering, out of the top pocket. Other pupils were gripped by this almost puppet-like trick and pestered me to make him do it over and over again. Forget pet mice or hamsters, forget gerbils, rabbits and guinea pigs, even as a very young weasel, Wilba received constant adulation. The other boys begged me to let them hold him, feed him, care for him, something I never allowed. I guarded him with my life.

I took him into class, placing his little box inside the desk in the days when school desks had lifting lids and round, two-and-a-half-inch diameter holes for inkwells in the right-hand corner. He would play happily inside the desk until he became bored and sought company. Then his pert little face would appear at the ink hole and he would skip out, run up my arm onto my shoulder and chitter excitedly into my ear. If a gowned master came stalking up the aisle I would thrust Wilba back down the hole and slide a book over it until danger had passed. We lived our days in mutual adoration, Wilba and I, his of dependence, mine in the grip of boyish captivation like a dream place, joyous, blessed, vividly alive, a whole new reason for being. At that moment in my careless youth all things weasel utterly consumed me. He was everything I had ever wanted.

Lives are often built on moments of joy and delight. Wilba and those school days are as vivid now as then, branded, locked in for life. I was a country boy and had developed an early interest in natural history, but how many country boys get to raise a weasel? How many get to box-trap mice and voles every day for a year? How many, at very close quarters, witness the stalk of live prey, watch the grip of dread freeze the rodent, wide-eyed, quivering with terror, unable to flee?

And then the quick, merciless dispatch, the dash, the pounce, the stabbing fangs and a sharp crunch at the base of the mouse's skull; the hot, limp corpse dragged away, a bead of blood on the sand.

★　　★　　★

19 March

Daylight lifts into a sky of shifting greys, flags of dense cloud billowing in barely visible motion like the sails of tall ships becalmed. It is calmer but still wet and chill, the hurly-burly of yesterday's return to winter temporarily on hold. The far hill on the south side of the glen stood black against a metal gleam rising from the east as I crossed the field. Light but not light enough. I dawdled.

Three red deer hinds eyed me warily from the field edge, heads high, long ears erect, motionless as bronze; they were top-lit effigies of the dawn, statues until they decided to go. An old hind, deeper in the belly, and two younger followers, perhaps yearlings. The old hind broke first and as she did so the low morning light flashed momentarily in her wide, wild eye. Six paces to the fence in single file, and up and over in an unhurried, effortless arc of instinctive grace . . . one . . . two . . . three . . . and silently away into the wood as though they had never been there at all. I crossed to where they had jumped and saw the dark imprint of their hind cleaves deep in the wet grass. I wished I could enter the wood like that. Caution climbing the fence, careful not to ping the wires.

Back at the wall, but now facing it, leaning against a birch trunk a dozen yards from where I had been before. Binoculars ready. First light is steeling the grey stones, the wet moss shining; the wren is somewhere behind me trilling out its piercing song-ditty, repeating over and over again with a volume utterly defying its size. I ponder what trick of evolution made the body so small and the voice so emphatically strident. How does so tiny a syrinx issue such penetrating volume – *fortissimo*?

Dawn enters the wood much later than the field. It has to prise the trees apart, shoulder its way in, so that I sit in gloom beneath bright twigs flaring at the top of the winter birches and pines. Weasels are both diurnal and crepuscular, darting shadows of the half-light. They go to bed in a nest of dry moss or grasses or sheep's wool harvested from fence barbs, curled in a tight ball, nose to tail, and, with a full belly, slide into a sleep the depth of oceans. Out of it. Night or day, only a predator can sleep like that. Mice, voles, rabbits and hares, even deer, can never risk closing down with such carefree abandon. Such are the lives of prey. Eyes maybe closed, but noses continually twitch and ears rotate like radar scanners. Every few minutes heads lift, wide eyes sweep, checking . . . checking . . . Drop your guard and you are doomed. Nature takes no prisoners.

So I wait. Daylight seeps into the wood. A hen black-bird alights too close to me, realises her mistake and departs noisily. Her raucous alarming echoes through the trees long after she has gone. A trail of fear issues like tracer fire. Every other creature knows that alarm and is instantly alert, an alertness spotlit by the first low sun scything in like a blade. The wren falls silent. I feel sleep tugging at my eyelids and have to stifle a yawn. Deep breaths and I am awake again. I glance at my watch: ten to seven. If there is a weasel in the wall it should show soon. For another half an hour, nothing.

Along the top of the wall, bounding from cope to cope, a weasel is heading towards me. It flows, unhurried, nonchalant, even sauntering. Its tiny feet seem barely to touch the stones, too fast to see, lighted and sprung. Still several yards away it rises on its haunches, fore paws dangling, little brown nose lifting and testing like a wine taster, white underside gleaming as bright as a snow-stripe, erect as a begging dog. The air between us fizzes with weasel inquisition. A stretched moment of risk assessment sparks inside its gyrating brain. It has seen me and knows what I am. I cannot fool this animal. The sharp little head momentarily swivels, drops and it bounds on and vanishes down the far side of the wall. Discretion has won.

An hour passes. I give up and slink quietly away. As I leave the wood I startle a buzzard perched high in a field-edge pine. It didn't see me coming from behind and I didn't see it until too late. I was close. The big hawk pitched clumsily into the air and flapped heavily five times before gliding away on stretched pinions. The sun flashed momentarily bronze on its back as it rose, then a black silhouette against an eye-squinting brightness of morning cloud. Its long, anguished cry echoed across the glen. Far out over the valley it soared on spread wings, turning in wide circles, winding higher and higher. I watched it fade to a speck, dwindling away to the far horizon until I could bear the brightness no longer.

20 March

I needed to try again. Earlier now, still dark and a heavy wetness of night rain laboured the air. From the long grass the redolent musk of red deer, rank and savoury, smote me in clouts of tart scent as I swished through. If they were still in the field I couldn't see them. Ribbed clots of fresh droppings gleamed claret-bottle green and shone like wet coal. I crossed the fence at the same place and stood beneath the buzzard's pine to see if my arrival had aroused anything in the wood. Nothing moved. The air was fungally rich with woodland fertility, a scent of wildness mingling with that of the deer and heightened expectation as if a fox or a badger might at any moment trot past. At times like this I wish I had the nose of a wild animal.

The scaly pine trunk felt welcoming to my touch. I paused there, needing more light. Minutes passed and a robin arrived. Pert, flicking from branch to branch, emerging from the twilight of the wood, it flew lightly to a birch beside the fence. Close. Close enough to see its beady eye, head tilted in my direction. It flicked its tail impatiently and its little fawn wings twitched anxiously. Perfect in every note, its precise, tinkling song clipped through the stillness, clear and urgent. Always the first bird to sing, it also knew that wherever humans tread, bugs and insects will reveal themselves. That ripple of song seemed aimed at me, seemed to demand action.

At the wall I took up my position against the birch trunk. Even as I did so I saw the white flash of bib as the weasel came out to see what the disturbance was. Its tiny eyes shone but it was still too dark to see form. A second later it was up onto the top of the wall and bounding away from me. Three days in a row was the confirmation I sought. I was now sure that within its depths the wall held a nest. A weasel nest. The first one I had stumbled across for years, and I guessed – no, fervently hoped – that it contained young. The robin had followed me in and was now singing lustily and the wren seemed to answer with its own trilling reveille. Dawn had arrived. The wood was suddenly awake.

Watching weasels is not easy. Even if you locate a nest, all you are likely to see is the female adult slipping in and out until the kits emerge for the first time. The male plays no part in raising young and may be nowhere near. What I wanted to witness most of all was the emergence of the young, but I had no idea of timings. Had the bitch given birth? When might the kits have been born? Was their emergence imminent, or still some days, even weeks away? Weasel kits don't leave the nest until they are four or five weeks old. I needed more data, more hard intelligence to inform a good guess. I had to keep watching.

21 March

A change of tactics. I arrived at midday, strode in carelessly and sat against another tree. The sky was vivid with the sun's equinoctial height, invisible behind a hard, squinting white of backlit nimbostratus. The wind and rain had passed over and left a grapefruit freshness in the drying air. The wood seemed to be lifting its spirits, shaking out its skirts, rooting for the close of winter. But spring comes late to the Highlands; false starts are the norm, like athletes jumping the gun and having to be hauled back.

High above, coal and great tits were sawing away, felling whole forests. A blue tit was busying about, to-ing and fro-ing in a hazel coppice to my left. It seemed to be gathering spider silk, threads streaming as it flew off. Their spring was under way whatever the weather. An hour slid by and the shadows of tree trunks edged across the leaf litter as the opaque, cloud-filtered sun soldiered west. A dunnock was exploring the base of the wall, ignoring me, systematically working the lowest crevices for spiders and bugs. I was wondering why it chose to stay on the ground when suddenly it flew, vanished in a straight line, away into a dense juniper bush.

The weasel came bouncing along the top of the wall with a wood mouse clenched in its jaws, unmistakable by the length of the dangling tail, normally held with such elegance and grace, now limp and lifeless. It disappeared down the far side of the wall. A few minutes later I rose, crossed the wall higher up and tiptoed down to see. There were so many potential entry points in those old stones, dark gaps and crevices between quilts of dark moss and fringes of lichen, that I couldn't tell which was its favoured portal. I needed to watch from that side, even though it was in shade. Ten minutes later the weasel's face appeared, saw me and vanished back inside the wall.

★   ★   ★

At four months old Wilba had grown into a fine, strong male weasel. He developed testes and I could feel a slender baculum beneath the fur of his belly. I established a routine. Every afternoon I took him out and let him explore the rough grass on the edge of the school playing fields. He would immediately fire into hunting mode, stalking through the grass forest like a tiger through jungle, suddenly freezing when he sensed a grasshopper and pouncing cat-like. I never saw him eat an insect, but I watched him crunch them and spit them out. It was as though he knew he was a predator and instinctively he was training himself to kill. Once he met a male stag beetle with a shining black carapace and fearsome mahogany mandibles protruding like spiky antlers. He stalked, paused, stalked closer and then, as if he couldn't stop himself, he pounced, veering away at the last second and immediately springing clear, seeming to know it might be hazardous. Then he bounded back to me, ran up my arm and chittered excitedly into my ear.

Every day I delivered a mouse or a vole from a ring of Longworth live traps – a necessary parental duty. If I caught too many I would release them again immediately. One a day was enough. To begin with I killed the mice with a finger flick to the skull, delivering them to Wilba hot and twitching. Even as a tiny kit he would launch himself at them, needle fangs crunching into the neck. Later I would free the hapless prey into his day box so that he could stalk them. Finally, as he approached adulthood, I would take him to rough grass, let him roam for a few minutes, watch him begin to hunt and then release the small mammal a few feet away. The pounce would be lightning fast and usually the terrified victim had no chance. Occasionally one would dart away and escape – wood mice were quickest and could sometimes leap clear – and Wilba, scenting the near miss, would hunt frantically through the grass, leaping and bounding, standing erect, dropping down and hunting on. When he returned to me and I held him to my cheek, I could feel his tiny heart racing like an engine until the adrenaline fizzed itself away.

When he killed he would drag the prey away into a corner to be out of sight and on his own. If I tried to take the food away he would shriek and chitter angrily, often refusing to let go. He would crunch open the cranium and consume the brain, on down the back, break into the chest cavity and devour the heart and lungs. Sometimes he would consume virtually the whole mouse, leaving only the jaws, feet, tail and shreds of fur. Sated, he would crawl off to his night box, curl into a ball and sleep the sleep of exhausted dogs. Out of it. When he awoke again hours later he would want to play, pestering me until I gave in.

Wilba was not a pet. I always knew the day would come when wildness and the need for a mate would claim him. It was a day I dreaded; a day I refused to discuss.

★　　★　　★

22 March

I headed out, uphill on the old farm track. In the avenue of ancient limes and horse chestnuts a fox surprised me – the briefest close encounter. It flowed briskly towards me in a nimble, bouncing trot, nose to the ground, intently pursuing some invisible tramway of scent. I was fingering the soft fur of goat willow catkins dangling in front of me. I froze, hand still in the air. The fox hadn't seen me. It came on.

A dozen yards away it stopped. It stood and stared. Uncertain. The peaked ears furred and rimmed in black, the swan-bright blaze of cheeks and chest, the slender black legs, the narrow, whiskered muzzle and the sharp, intelligent face. Our eyes met – and held – just long enough for me to read the dread in those pale amber orbs, a miasma of terror instantly shedding into the morning like thistledown caught in the breeze. This is sheep country. Foxes kill newborn lambs. Round here farmers and crofters shoot them on sight. This one was taking no chances. It turned and fled, instinctively deploying its skilful use of cover, first into a thicket of broom and gorse, then out of sight along a ditch, not emerging again for sixty yards. It broke out into a field and streaked across open ground towards the far sanctuary of woodland, scattering sheep as it ran. Its bushy tail streamed like a pennant in its own slipstream. The sheep bunched in panicky clots to watch it go. Long after it had disappeared from view its uric, peppery scent lingered among the avenue trees. It was eight o'clock. For those few seconds of locked eyes, that fox had burned furnace red, the red of rowan berries before they fully ripen and the late summer rouge of old English apples – a glowing fusion of oranges and reds, all impaled on a beam of low morning sun.

A dawn mist had held me back, stopped me heading out earlier. When at last the sun burned through, the universal legacy of dampness was everywhere – not the dripping wetness of rain, but the bright, breathy moisture of spray, ethereal, side-lit and blown. Cobwebs shone with droplets like seed pearls on Elizabethan lace. Frondose lichens bearding the birch twigs looked glazed in viridescent icing, light spun back from their usual powdery surfaces.

The mist still filled the valley and out over the river. As far as I could see the glen below me was a lake of polished pewter, its surface gently undulating on a faint breeze funnelled by the conduit of the glacial valley. As I watched it began to melt away in the lifting sun. There was no morning haze; a needle-sharp clarity had claimed the land, pellucid and still. Above the lake the far hills looked strangely new, scoured and burnished by the wind. Rags of bright cloud broke and faded away, exposing a savage, overpowering blue. The air was light and seemed to carry me up the farm track towards the wood.

At the wall I sat in a cosmos of birdsong. Freshness glinted from everything the fierce sun touched. It had aroused the whole wood, which seemed to vibrate with competing strains, like an orchestra tuning up. Cock chaffinches were shrill with self-importance, prominently perched, puffing themselves up, repeating over and over again. Woodpigeons crooned distantly. Somewhere off to my left a greater spotted woodpecker Morsed out its ringing drumbeat. Dunnocks jingled mechanically from the undergrowth. An anxious mistle thrush prattled peevishly and invisible wrens trilled insistently, 'Sounds and sweet airs that give delight and hurt not.' And I, Caliban, mute as a bollard, the monster in their midst, happy that they chose to ignore me.

Since I now knew the weasel was nesting there, a sighting of some sort was almost guaranteed. All it required was patience, a shutting-down patience retreating into my head, letting the wood and the birdsong and the sunlight flood over me like a warm tide and draw me in. There are occasions when separate strands of nature come close to a perfection of refinement, even though they are random and unintended. This was one of those moments: the vibrant birdsong, sunlight dappling through branches, the bright air redolent with spring greening, all gift-wrapped in the illusion of tranquility. I could feel my spirits rising, spiralling skywards, entirely happy to sit and wait.

I had yet to determine which of the dozens of caves and crevices at every level in that wall would be the regular weasel entrance, if indeed there was just one, but it proved harder than expected. If you stare at a dry-stone wall for long enough, it befuddles you. You clock movement that isn't there. Your mind whirls. You decide on one dark crevice and then lose it, forgetting which one it was. The wall blurs itself into your consciousness like a repeating pattern on wallpaper. A movement alerts you then disappoints when it is just the light-tremor of shifting leaves high above. Minutes crawled into an hour. I stared and stared.

I tried looking away, at identifying the plants at my feet: the emerging clover-like leaves of wood sorrel, the delicate pink tracery of sphagnum mosses, the winter-killed wavy-hair grass in pallid clusters, the square-ribbed stems of bilberry budding up after winter. I picked up larch twigs wind-ripped by the winter gales and marvelled at the tiny bundles of needles at each node, stars impaled on a skewer. Then I stared at the dark mosses and pale lichens patterning the boulders with shimmering slug trails inscribing them in hieroglyphics, a secret coded text, random yet precise in their intriguing ramifications.

What seemed like an age might not have been that long, perhaps an hour and a half at most. Wildlife watching always demands patience. But instead of the weasel bitch emerging from the wall, to my total surprise she appeared from behind me, darting past the tree I was leaning against and bounding on towards the wall. At first I thought she hadn't seen me, but before disappearing into a crevice she turned, rose up in begging mode and looked straight at me. I held my breath. Raised teats stood out on the white of her belly – the confirmation I sought. She was checking me out, but no real fear showed in her little jet eyes. A shared feeling of identification seemed to fizz between us. Could it possibly be that this bright little animal, this tiny twist of fruitwood fur, this scrap of carnivorous wildness, after only a few days had begun to accept me and my unavoidable man-stink as harmless?

In that briefest moment of proximity it felt as though an unspoken acknowledgement had passed between us, a visceral, silent, animal reconciliation between two sentient beings. She vanished into the wall and I logged her entry, three courses down from a confection of moss, spinach-green, spilling down from the copestones like a child's icing on a cake. An hour later I gave up and left the wood. There are times when very little is enough.

23 March

I slept badly. There was a weasel in that wall and she had kits. Fact. It was decades since I'd seen a weasel kit and the certainty was tugging me back. Sooner or later they had to emerge and I wanted to be there to witness it, but I knew the odds were stacked against me. The challenge nagged at me all night.

Every spring I find birds' nests and check them out daily to try to see that testing, often comical moment when the fledglings pitch into the air for the first time. Sparrows, robins, swallows, house martins, wrens, spotted flycatchers, pied wagtails, chaffinches, dunnocks, jackdaws, thrushes and blackbirds, all the busy, commonplace denizens of the garden habitats in and around the house that bring so much joy to our spring days. Occasionally I time it right; mostly I miss. But it never fails to cheer me to see new life lift off, even though those lives are often brief and end with a snatch from the darting sparrowhawk.

At dawn I hurried back across the field. Weighty clouds, as grim as a wintry ocean, were barrelling in from behind the mountains in the west, mingling with layered ranks of subtly differing shades to the east. Where they collided with first light, they pinned it down to a spreading gloom so that when I arrived at the wood all yesterday's optimistic glamour had disappeared. Without the sun the air was brazen, a harsh, clawing dampness without rain. I crept in with burglar stealth, further down the wall than before, much lower, hoping to contain my scent away from the nest, treading the sodden leaves of winter, avoiding the tangles of wind-strewn twigs beneath the pines and larches as I picked my way uphill.

It is folly to imagine a man can enter a wood undetected. A red squirrel came scrabbling vertically down a pine trunk, disappeared round the back and reappeared at my eye level, peering round to check out the intruder. It bounced out onto a branch, stopped in full view, eyeing me sideways. Its round dark eye was ringed in pale fur. Its long fluffy tail hugged the curvature of its back with all the insouciance of a cavalier cloak slung over a shoulder. The staring eye and extravagant ear tufts of winter lent it a combative tenor, a doughty demeanour both tetchy and charged with inquisition, as though I had somehow invaded its critical distance and caused it offence. I stood my ground. A great tit arrived above the squirrel and alarmed loudly, a sharp, monosyllabic mocking refrain repeating over and over again, shrill and vehement, never still, never silent. The squirrel flicked its bottlebrush tail and skittered off up the pine. I could hear its claws on the scaly bark long after it had gone from my view.

I took my usual seat facing the wall. An hour passed, then two. Wood mice came and went, skipping nervously over the woodland floor, darting in and out of the base of tree roots, foraging among the leaf litter. Their long, wire-like tails flourished out behind them with all the sprung elegance of calligraphy script, flowing, never touching the ground.

A few minutes later a medium-sized brown rat emerged from the wall some way off to my left. Entirely unaware of my presence it came sniffing along the ground at the base of the wall, its long snout twitching. I guessed it was hunting the wood mice. Rats are often carnivorous, well capable of catching and devouring mice and voles. This one had made a fatal mistake. In a flash of russet fur the bitch weasel shot out of the wall and grabbed the rat by the neck. It shrieked, a thin, anguished shriek cut off and stilled by a crunching bite to the back of the skull. Although the rat was slightly bigger than the weasel, it only took a few seconds to drag the dead rat into the wall between basal boulders. This was the first time I had witnessed a weasel kill for many decades. I had forgotten just how devastatingly fast, how vicious and fierce the attack was, and how lethal.

It took two seconds to kill that rat. The rodent never had a chance. The weasel must have detected the presence of the rat while it was still a little way off, positioned itself inside the wall to affect an ambush, and waited poised like a mainspring. It attacked from above, perhaps six or eight inches, springing down and landing squarely on the rat's back. The bite was instantaneous, time only to utter that squeal of shock, of fright, of terminal alarm.

Watching that dramatic act of predation brought all the memories of Wilba flooding back. All those years ago I had seen him kill so many times, hundreds in fact, that witnessing it again carried with it no sense of surprise, no sudden revelation or 'Wow!' moment – just an unstoppable tsunami of vivid déjà vu memories winging in and surrounding me in a gentle cushion of reflection.

Fired up by this sudden surge of unexpected action, I decided to stay longer. I leant back against the tree. Wholly absorbed by memories, a deep sleep slid in and took me over.

A spitting, windless rain awoke me, stiff and cold. I was horrified to see that nearly an hour had passed. While I slept, bruising clouds from the west had closed ranks and a universal dankness had sprawled throughout the wood. I decided to pull out and head home. There are limits to the discomfort that even dedicated naturalists will tolerate. As I moved to get up and stretch, white rump flashing like a beacon, a jay screeched in and landed on the copestones in front of me, a colourful and rowdy incursion into a dank woodland morning. Its grating call echoed through the trees as it hopped from stone to stone and an answer rasped back from far away on the other side of the wood. It seemed to detect my presence and leapt into the air on broad wings. It vanished.

Guessing that the weasel had probably left the wall, and wondering if I could detect any weasel scent, I moved close and bent to the triangular crevice where she had entered with the rat. As I did so a faint sound emerged, so faint that at first I thought I had imagined it. Then it came again, not just one but a scratchy fusion of several tiny voices. My heart leapt. Not from that basal cavity, but from another, higher up and further to the left, I heard the faint but unmistakable, unforgettable – so utterly unforgettable – bleat of weasel kits, a sound I had not heard for decades, a sound branded since boyhood. Thin cries, as thin as bible tissue, piteous, straining, pleading, oh-so-tiny voices of hunger and hope.

Not only did I now know where they were, but I also had a fair idea of their age – certainly not yet weaned. They must have heard me, detected the indistinct rustle of my presence and mistaken it for their mother returning to suckle. It was a moment of entirely accidental, slightly shocking elation, as if two secret lovers had rounded a corner in a supermarket and met unexpectedly face to face, smiling and shaking their heads, not knowing what to say. If I hadn't been bending so close, ear to the wall, I wouldn't have heard it at all.

★   ★   ★

That was it. That was the sound I had heard that day so long ago when I found my weasel. It was the cry he made for the first three weeks while I fed him droplets of milk from a pipette, the cry that awoke me in the night when his box was on the pillow beside my face. It was the cry that slowly matured into the stronger, firmer, more persistent chittering calls of adolescence, sounds I remember so well, sounds inseparably bonded with his musky scent, both wired in for life.

As he grew his coat rapidly changed from a dull, tawny lion to a longer, brighter and sleeker fur of henna russet. From his chin to his underbelly he glowed a clean, shining white. His milk teeth disappeared and were replaced by fangs, ferocious weapons for ripping and killing, curved canines as lethal as a falcon's talons. He discovered how to lap milk from a saucer. And he was faultlessly clean, both urinating and defecating in a corner of his choosing and groom-licking his fur with an almost obsessive dedication.

When he had fed and slept he needed to play. I made him a big box from an old cabin trunk with a glass lid. Weasel play is boundless and relentless; the animal seems sprung, fired with an elastic energy as fast and agile as the wood mice he has to be able to catch. An obstacle course of tubes and tunnels, hidey-holes and cardboard boxes taped together kept him happily engaged for hours on end. His curiosity was endless and uncontrollable, so typical of many members of the weasel family. He would not rest until he had explored everything within his realm. A woollen pom-pom hanging on a string from the box lid seemed to arouse all his killer instincts and transported him into squeals of ecstasy. If it swung like a pendulum he would impale it with piercing jet eyes, whole head following the swing, and then in a streak of Titian fur, spitting and hissing fury and defiance, fling himself at it full tilt, sinking his teeth into it with mock rage, swinging backwards and forwards as he fought hopelessly to kill it. When he finally gave up he would vanish into his small nest box and plunge into his weasel sleep-world as dark as a tomb, until hunger aroused him once more.

A special delight was half-filled bookshelves where he could weave in and out of books, behind, on top of and between every book he could reach, tipping them over so that they fell flat and then nosing them off onto the floor. Peering over the edge of a shelf he would spy another ten or twelve inches below. In one flowing acrobatic manoeuvre he would swing down and land on all four feet on the lower shelf to begin his literary rummaging all over again. In all the hundreds of hours of rough play he never once bit me in anger. He would often grip my fingers in his tiny teeth, or even sometimes my ear lobes, but never enough to draw blood. He seemed to know precisely the acceptable pressure for play, a world apart from the merciless attacks on his prey.

★   ★   ★

24 March

Rain. Wind and rain all night and all day. One of those days when driving rain overwhelms the sky, the mountains and moors, the fields and woods, the very soil itself.

Horizons vanish so that we feel imprisoned, locked into a confined, turbulent world. Days when it is better to crawl back to bed rather than venture outside. Days so unforgiving that an empty hopelessness corrodes the frayed edges of your soul. I went out as far as the field, got buffeted by brutal gusts, saw nothing but wind devils whirling twigs and soggy leaves into corners against the fences and buildings. Within an hour I came home, drenched and shivering. In the distance I could hear branches snapping and whole trees crashing down in the spruce plantation. The storm rampaged throughout the day without pause. Rain drilled down in monsoon-like torrents. Sitting beside the fire later that afternoon I wondered how weasels fare – how any wildlife fares – on days like this.

25 March

A roar of angry water dragged me from sleep. The cyclone had seethed through the Highlands for thirty-six hours, the last of the equinoctial gales from the Atlantic. The dawn handed us back the hills washed with streaks of brilliant vermilion, purpling at the edges, the last gasp of the departing storm. At sunrise those hills were still on fire, the reds bleeding away to yellows against a shining blue so bright that I wondered whether it could ever be grey again. Venturing out at ten, the saturated lawn steamed wetly in the sun.

From the overflowing loch the burn was rampaging down to the river, white with splenetic foam and spume, a proper spring spate defined by the dull rumble of boulders being bowled along the bed of the stream. I went to see. There is something humbling about angry water. A humbling touched with real fear. I could feel the mist of spray on my face and lips. You know in your bowels that if you fell in you would die. This was not water simply flowing fast, but raging through, tumbling over, under, around and between banks barely containing its ire. Instinctively you step back. Overhanging branches were being ripped off; whole chunks of bank soil gouged out and vanishing in a second, gobbled by the hurtling stream. Your eyes can't focus on water passing as fast as a train. You come away charged and giddy with the force of nature.

Half a mile down the river where the Aigas Dam truncates the gorge, the hydro boys had opened the emergency spillway to stem the valley floods upstream. A mini Niagara of cappuccino-coloured water crashed furiously into black rocks, the morning air fogged with spray and the insipid reek of dank caves. I hoped the otters had moved their cubs.

At ten-thirty I crossed the field. The Highland cattle also steamed. Their shaggy winter coats had shed most of the wet, but they'd had the sense to move into the sun and settle down to ruminate. Where the sunlight struck their backs they also appeared to be on fire, smouldering gently. Their deep amber eyes beneath each huge spread of upturned horns gave them a sinister air of foreboding. If I didn't know how gentle and docile they were, I might have thought them fearsome other-worldly beasts, like the forest aurochs of a bygone epoch.

The wood squelched underfoot. As I walked up the wall, little rivulets of rainfall streamed off the high knolls, washing last year's leaves and tiny larch cones into allur-ing patterns of miniature deltas and alluvial fans sep-arated by long, thin islets of woodland debris. I worried. Had the wall been breached by the deluge? Was the nest safe? Was it still dry? I knew that most mammals will move their young to safety if they feel threatened. I also knew from those thin cries and the sight of the bitch's teats that the baby weasels were nowhere near weaned. Between two and three weeks old seemed likely. Kits wean at four weeks and are eating meat and learning to hunt with their mother at five to six weeks.

I checked out the wall well away from the nest site. In a gap big enough to squeeze my hand in among the stones two courses down, I found it chill, but not wet. Although the tight-packed fronds of the moss cushions on the copestones held moisture, their root-mats were dense and dry, like a turf roof. Where I reckoned the nest to be was well covered with moss. Relief. I was sure they would still be there. The sun had stirred the birds again and, although still dripping, the wood rang with a spontaneous orchestration of exuberant song. The sound flooded over me, dispelling all misgivings. I left the wood with a spring in my step.

On my way back across the field I stood to watch a molehill erupt. Nothing visible above the surface at all, but the fresh brown earth humped up from below in sudden convulsive eruptions, a mini volcano, three or four pulses in quick succession and then a long pause before another. The rain had brought the earthworms close to the surface, easy pickings for moles excavating new tunnels with powerful shovel forearms hydraulically strong, shunting the spoil to the surface in a long chain of crumbly brown tumuli.

*   *   *

The day came. I had to steel myself; a sickening dread churned in the restless crypts of my gut. Wilba was in my jacket pocket, curled, content, sleepy. With a school friend I walked slowly and quietly to the woods, to the ivy-covered log where he was born and I had first found him ten months before. In a plastic bag we carried a live field vole. I lifted Wilba out in my cupped hand. He sat up and sniffed the autumn air, ran up my arm and sat on my shoulder. Then he ran down again and bounced out onto the woodland floor, confidently expecting the usual release of prey. I placed his small sleeping box under the log, still stuffed with weasel-smelling cotton wool, so that he would have somewhere to go for the night if he needed it. We sat and watched him musteling about through the leaf litter and the dense ground cover of ivy, probing, prying, searching. He was fully alert, eyes bright, blackcurrant nose constantly testing. An adult weasel. He bounded back to me, ran up my leg and sat on my lap, chittering with impatience and hunger.

Slowly and carefully we opened the plastic bag and eased the vole out onto the ground. It sat quivering for perhaps three seconds before scuttling off into the undergrowth. Then I placed Wilba gently onto its trail. His whole demeanour changed. He darted forward, back and then off again, instantly sparked into full hunting alert. The ivy rippled like the path of a trout in a shallow stream, back and forward, to left and right . . . a faint, thin shriek told us that he had caught the vole. Then we saw him emerge, the dead rodent limp in his jaws. He ran to the log and disappeared underneath it, vanishing behind its dark ivy curtain.

In a funk of planned and rehearsed opportunity, we got up quickly and walked away.

Did I look back? I can't remember, but I can remember the silence and the deeply burning loss that engulfed me as we hurried home. Please do not tell me I am sentimental. Decisions are taken and judgements made, primed by the emotional demands of the moment. But the truth is the truth and I cannot truthfully recall the final departure of that little animal without also revealing the emptiness and the hollowing out of everything I had felt and revelled in for many months. I was sixteen, an emotionally convoluted moment in any young life. That weasel had wholly captivated me by night and day, week in and week out, but I was old enough and sufficiently nature-wise to know that for all his endearing qualities, he was not a pet – there was not one whisker of domestication on him – he was wild and the incorruptible essence of wildness pulsed through his veins with a fierce and relentless tyranny. To the wild he belonged, not to me, and it was to the lottery of the wilds that he had to return. Did I go back? Yes, I did, the following day. I stood at the log for ten minutes, gritting my teeth to stop myself calling his name. I saw nothing, heard nothing and felt nothing but the overpowering burden of my own desolation.

But that was not quite the end of my weasel story. I was well aware that returning hand-reared animals to the wild is a deeply uncertain and risky undertaking, often leading to starvation or death by predators of which the animal has no experience. So many out there for an unwary weasel: buzzards, crows, badgers, foxes, cats wild and feral, barn and tawny owls, kestrels, hen harriers, even herons and gulls have been seen to stab at them. Here in the Highlands weasel skulls have turned up in golden eagle pellets. So many perils, threats and uncertainties. How could he know about predators? Could he hunt well enough? Would he shelter and sleep in his box? For how long before he found and made his own den? Would his absence of fear of humans lead to an early death? Had I done enough? Would he find a mate? These churning worries kept me awake at night and many times I thought of trying to get him back – but I didn't.

Almost exactly a year later, on a high summer's day of sun and the heady fragrance of freshly mown hay assaulting our nostrils from adjacent fields, I was strolling through those familiar woods with my father. We passed by the log. 'I released Wilba just here,' I told him, pointing to the ivy-clad elm. We walked on in silence. A few minutes later, only *a hundred yards or so away, we sat on a sunny bank to eat our sandwiches.*

We had been there quietly talking for about half an hour when a fine adult weasel in a glossy chestnut coat suddenly appeared out of the long grass, sat erect in begging stance and stared straight at us. I held my breath and my heart seemed to stop. The weasel dropped and bounded straight to me, ran up my leg and onto my knee. We stared at each other for a fleeting second before he turned and skipped away into the undergrowth. That was the last time I saw Wilba, but my heart sang with a deep and thankful gladness. He had survived, wild and free.

★   ★   ★

8 April

Two weeks later it felt like spring. Wind from the southwest as soft as lamb's fleece, marking the new season like a gift. The warm air of sunrise towered away to high, radiant clouds beating slowly from beyond the mountains. The birds rejoiced at dawn, settling back into a steady, echoing chorale when I first stepped out. The year's first butterfly skipped lightly over the field with a vivid brightness of pale wings and apricot flashes flickering like Belisha beacons – a male orange-tip searching for the duller, soot-tipped females. They were out quartering the meadow for the pink-flowered lady's smock, laying just one bright orange pinhead egg on each plant.

High above, silhouetted in the squinting blue, a chevron of seven whooper swans rowed majestically north. The whaup of their wings and their softly fluting whoops lingered long after they had passed from view. They were migrating back to their Arctic breeding grounds in northern Scandinavia. The birdsong, the butterflies and the swans seemed like a validation and an augury for everything good. The Earth was spinning back towards the sun on its long ellipse, as it should, as it must, the season opening up once again, fighting free from the long dark days of winter.

Watching and listening intermittently, I had seen little but heard the kits' cries slowly growing stronger and more demanding. Every day the weasel bitch was bringing in prey – voles, mice and occasionally a shrew – sometimes in the morning just after dawn, often in the evening as the light was fading. Once I was too close with my ear to the wall when she came bounding along the copestones with a vole in her jaws. She dropped it on the top of the wall and fled. The vole was still warm. I backed off. Ten minutes later she returned, snatched it up and darted down into the wall. Against the birch tree she ignored me.

10 April

Dawn crept in like a thief. A tawny owl, silent as a falling leaf, floated onto the copestones, peered around for a few seconds, logged me with its dark scotopic orbs, glared rancorously, unamused, and lifted off again, springing into the morning air and ghosting away through the trees as though it had never been there at all. If I knew about the weasels, it did too. Predators survive on the folly of the unwary. That's the life of a tawny owl: it sits and watches and remembers. Movement and sounds are meticulously recorded for action or for future reference. Its brain holds the bounds of its hunting range inscribed like an annotated map. Regular perches are logged in order of hunting success: excellent, good and indifferent. Owls never forget.

Oh! I was slow, so very slow and so blinded by my own mistaken certainties. I should have read the runes – they were clear enough. The owl on the wall should have screamed out to me. The weasel kits were out and the ever-watchful owl knew it, had charted their movement – had probably been eyeing up the weasel nest for some time – hoping. A weasel kit on its first outing would be a tasty meal for a hungry owl, easy prey. I was too dull, too blinkered, too wrapped in my own nostalgic watching world, too apart and unaware, altogether far too human. Our world is an 'other world', a world apart and a nowhere-near-parallel universe. We left the woodland world of weasels and owls aeons ago and we cannot return.

I waited in the warm, resinous twilight of the pine trees. An hour, then two. Columns of sunlight floated in. Shadows pooled, seemed to darken then gently bled away, the air heavy with unseen presences. Motes of woodland dust, charged by the gentlest of breezes, danced in sunbeams as if they were gnats. Slowly the wood awoke, came alive; from somewhere off to my right a jay's demented screech tore through the silence like an alarm. Birdsong responded in a rising chorus, framing the wood in joyous bursts. A dunnock flickered onto the top of the wall in front of me and vanished again.

As sharp as tintacks, the two hoodie crows had watched me leave the field; they came to check me out. They levered past and landed out of sight, rasping loudly from on high. For ten minutes they sat where the buzzard had, in the dead and twisted antler-top of the old pine, hoping for action. Eventually they gave up and winged away over the valley, paired like sinister shadows. Their departing calls grated mechanically through the branches.

For an hour the wren had been singing almost continuously. Its melodious trilling was all around me, vibrant, insistent, filling space, proclaiming and fencing territory with a force and volume absurdly disproportionate to its size. It out-sang the robin and the chaffinches, the tits and the creeping dunnock, louder even than the warbling blackbird and the scratchily fluting mistle thrush. It trilled and trilled. Through binoculars I could see its tiny bill tilted to the sky, amber throat spread, tail vertical and wings oscillating from the shoulders in time with the trills. It hopped from twig to branch and back again, changed direction, challenged louder and louder with a vehement, assertive defiance. And then it stopped.

Still I didn't get it. The wren flitted to the wall in a blur of tiny wings, landed on a moss cushion and grated angrily like a cricket. A second later it was gone, back into the undergrowth, still prattling alarm, relentless and frantic. Slowly I began to realise what was happening. I sat up, suddenly wide awake.

They came so fast and so suddenly that I didn't know where to look. From the ageless silence of stone and moss, the dyke was suddenly writhing with weasels. They came running, jumping, bounding, scrambling, dipping and diving in and out of crevices, leaping, squirming, leap-frogging and ambushing each other, in and out again, over the copestones, down to the leaf-strewn ground, up again, vanishing, popping out, ludicrous and frenzied like a mad, crazed puppet show. It was a dazzle of weasels, a weasel spectacular, a mustelid fantasia, an exhibition of nivalean élan, uncountable, insatiable, unstoppable and utterly spellbinding. They were a riot of beautiful nut-brown vandals on the rampage. Over and over again I failed to be able to count them. There must have been five, six or even seven kits, discernible only by their blunt noses and with their mother, a ravel of weasels, of wriggling, writhing, furry trapeze artists, rolling, tangling, pouncing, darting and skipping with a boundless elastic energy, so quick and so breathtaking that when they disappeared as fast as they had arrived, I was exhausted.

I didn't see where they went. Back into the depths of the wall? Perhaps. Or down the other side and away into the wood in an affray of demonic hunters like a supercharged posse of musteline hounds? I shall never know. I waited another twenty minutes, praying for them to come back, but – nothing. Slowly and reluctantly I got up and left the wood.

As I dawdled back across the field my mind was fired with weasels – the weasels I had just witnessed and hot, resounding echoes of a weasel long ago. I hadn't seen them emerge; they must have been out for some time. There had been no hint of first flight, no suggestion of caution or of tentative venturing out into a dangerous world. That weasel mob were joyful, play-practised youngsters who knew their surroundings with an easy, flamboyant and brazen familiarity. I think they had been out learning with their mother for at least two days, maybe more, perhaps hunting in a pack, honing skills, foraging, prying and probing in their inimitable weasel way and, hopefully, doing what weasels are designed for, finding and killing prey.

3

Badger

The badger grunting on his woodland track
With shaggy hide and sharp nose scrowed
   with black
Roots in the bushes and the woods, and makes
A great high burrow in the ferns and brakes.

'The Badger',

John Clare (1793–1864)

Family *Mustelidae*, the weasel family of the order *Carnivora*. Subfamily *Melinae*: the European badgers. There are eleven species of badgers worldwide, divided into several different genera. The British badger is *Meles meles* from the Latin *melum*, meaning a badger. French *blaireau*, German *dachs*, Spanish *tejón*. The English name 'badger' is derived from the Old French *bêcher*, to dig, *bêcheur* a digger. They occur throughout Britain but are absent from most offshore islands. Badgers are also widely distributed throughout Europe.

The badger is a stocky, short-legged, shaggy-grey-coated, bear-like mammal weighing 8–14kg, with the defining feature of two prominent black stripes running through the ears and eyes to the nose and the length of an otherwise white face. Zoologists label this 'aposematic' colouration, stripes and colours as a warning signal, such as on wasps, skunks and many poisonous snakes. In badgers, the prominent striped face acts as an alert to potential predators that it has a savage bite, a short temper, and is well able to defend itself. Its skin is thick and body coat consists of long grey-tipped and black-banded guard hairs over a shorter grey underfur.

Males are called boars, females are sows and the young are cubs. Badgers have short, rounded, white-rimmed ears and small, rather piggy eyes positioned in the black facial stripe. Eyesight is very poor, but hearing is acute and their sense of smell is exceptional. They are prodigious diggers, excavating deep, many-chambered burrows called setts, often on a slope or in a bank. Ancient setts may have many entrances and tunnels linked underground covering a wide area, easily identified by the huge heaps of sandy or chalky spoil outside the holes. They are largely nocturnal, emerging at dusk and spending the dark hours foraging over a wide range.

Badger habitat is typically in woods and copses, field margins and nowadays in suburban gardens. They occur from sea level to high in mountains. They require soft or sandy soil to dig their setts, and moist soils for *Lumbricus* earthworms, an essential component of their diet. They are very territorial and will fight fiercely to evict marauding badgers from neighbouring territories. They are capable of living in large colonies of five to fifteen animals or much smaller groups, and even as solitary animals, depending upon the available food supply. They are also very clean: they collect grass or bracken for bedding, dragging it backwards in a bolus under their chins and down into their sleeping chambers. They regularly eject soiled bedding from their setts and replace it with fresh vegetation. They dig shallow latrines or dung pits in the vicinity of the setts and elsewhere to mark territorial boundaries.

In common with other mustelids, badgers have a carnivorous dentition with well-defined canines and powerful jaws, but back teeth are adapted for an omnivorous diet. The upper and lower molars are flattened for grinding plant material: cereals, fruits, nuts, roots and tubers as well as beetles, grubs and, most importantly, earthworms, which constitute up to 70 per cent of their diet. They also catch and eat small mammals, ground-nesting birds and their eggs, nests of young rabbits, amphibians, hedgehogs and some carrion. A great favourite is bee and wasps' nests, comb, eggs, grubs and adults, all dug out of the ground with powerful claws and greedily consumed, apparently oblivious to stings.

By heavy foraging, both sexes need to put on brown fat in the autumn of the year in order to be able to survive through the winter months of snow, frozen ground and very limited food. In bad winter weather badgers often don't emerge from their setts for several days at a time, but they do not hibernate.

As with some other members of *Mustelidae*, badgers possess delayed implantation, the genetic adaptation which allows the fertilised ova to be retained in the fallopian tubes until the sow badger is in sufficiently good condition for the blastocysts to embed in the uterine wall and begin developing. This enables mating to take place in any month of the year, with birth generally in February or March after a gestation period of seven weeks. Two to four cubs are born blind and pink with soft silver fur. Their eyes open at four weeks and they emerge above ground for the first time at around eight weeks. Weaning starts at twelve weeks, although they may suckle until five months old. Cubs can be very playful, with much scampering and leap-frogging of each other.

Today they have virtually no predators other than humans, although in historical times unwary cubs may have been taken by wolves, lynxes, brown bears, wolverines and even eagles; nowadays the greatest threats are direct human persecution and roadkill. Up to 50 per cent of the badger population of Britain is killed every year on our roads. Most people have only ever seen a badger dead at the roadside. They have been protected in British law since 1973, but in some country districts badgers are still illegally dug out for the savagely cruel, so-called 'sport' of baiting with fighting dogs and betting on how long the wretched badger can survive. Its death and often severe injury to dogs is the inevitable outcome. This vile practice continues today in secret.

Badgers are vectors for bovine tuberculosis, a disease accidentally introduced into Britain from the Continent by cattle farmers in the 1970s. As a consequence badgers have been charged with spreading the disease and, despite being a protected species, they have been widely persecuted, both officially and illegally, in dairy- and beef-rearing farmland in England and Wales. Trapping, shooting and gassing have been the principal means, while illegal snaring is also often practised. Badgers are easy to snare because they habitually use regular paths to and from their setts and feeding grounds. There is no clear scientific evidence that badgers are solely responsible for spreading the disease – there are several other vectors such as fallow deer and hedgehogs – but the blame sticks and successive governments have authorised controversial localised badger culls.

★ ★ ★

Here at Aigas, badgers are with us all the time. They visit the field centre's wildlife hides every night, drawn in by peanuts and a smear of honey. Our guests love seeing them close up and the badgers are always happy to oblige.

For me, having lived with badgers all my life, they are fixtures, as much a part of the living landscape as the pines and birches, the spring orchids on the moor or the autumn-fruiting fungi. The land would not be the same without them. Aigas is their habitat, and that suits them just fine. On my daily walks, with or without the dogs, I see their presence everywhere I go. With practice you get your eye in. Unmistakable – a rootling here and a scrape there. The broad, slightly flat-footed paw prints in soft ground, the jigsaw-piece-shaped sole and five toe-pads just registering the long front claws as arrowhead dimples in front of each digit, and usually the hind footprint clipping it behind. Their habitual nightly foraging rounds can cover several miles. You get to know their routines, their paths well tamped down in a way that a deer or sheep trail wouldn't be, winding through the woods. And then, every now and again, you stop and stare. You can't quite believe your eyes. All hell has been let loose.

When a badger finds a wasps' nest within its reach, perhaps in the stump of an old tree, or a mouse or vole hole leading into a root cavity below ground, even in among the basal boulders of an ancient dry-stone wall, the full force of badger power is brought to bear.

There should be a word formally to acknowledge the physical strength of badgers: perhaps 'badgerforce' or maybe just change the meaning of the word 'badgered' to mean not just harassed, but totally wrecked.

You have to witness it to understand the forces at play: boulders hauled out of the ground and shoved aside, roots ripped and torn, earth and stones flung wide, and a large crater of nest excavation, the whole-sale scattering of those delicate paper walls and cells so intricately and precisely mouthed and fashioned in shell-like arcs, litter strewn in all directions. And the wasps? Gone, gobbled – every last one. Sometimes, in the sobering chill of morning, I have found one or two bewildered workers hovering around, no doubt wondering what species of monster has totally obliter-ated their home. A deathly silence has descended. It looks as if something mechanical and hydraulic has been at work. All the grubs guzzled, the hundreds of egg cells simply vanished, the scores of worker wasps and all the developing queens utterly hoovered up, entirely regardless of stings. The angrier the wasps became and the more they attempted to attack the badger, the more they got munched – every last one that it could find, before finally shuffling off on its rounds, leaving a scene of wholesale devastation. I don't know whether anyone has ever witnessed such an event, far less caught it on film, but on the list of British nature's wild flings I would like to see, it is definitely in the top ten.

A good friend, a retired scientist who was a badger researcher in Devon, experimented with weights. He hid peanuts under a steel dustbin lid and then piled heavy stones on top of the lid. His badgers were not the least bit fazed. Night after night they hauled the stones off, flipped up the bin lid and guzzled the nuts. Then he tried concrete blocks, first one or two, then doubling and trebling them. Same result: blocks simply hauled off and shoved aside. Finally he pegged the bin lid firmly to the ground with steel spikes, then heaped boulders on top. The badgers seemed to be mocking him. It was as if they were saying 'Wake up, pal! We can match all your tricks.' They rose to the challenge by digging a tunnel under the lid, leaving all the boulders in place. Game, set and match to the badgers.

In the woods the badgers are performing a valuable ecological function. Ecologists have labelled them ecological engineers, a species which alters the natural environment for themselves and benefits other species at the same time. By digging setts they also provide homes for foxes and many invertebrates; they turn over nutrients and they create seedbeds wherever they go. Whether it is just a little hollow in soil where an earthworm or a grub has been rootled out, or whether it is a bigger hole such as where a bumblebee nest has been excavated, a bare earth declivity is left behind. An open invitation to seeds – any seeds – floating by. And then there are elder trees. It is too much of a coincidence that many – perhaps even most – badger setts have elder trees beside them. Badgers love elder-berries. When the berries begin to rot and fall to the ground, badgers enjoy a mildly alcoholic feast. Then they toddle off back to their setts and defecate tidily in carefully dug latrines close by. Richly fertilized, the seeds germinate and hey presto! You have an elder tree. At one site I know the setts are ringed in by elder trees. I've never done the research, but it must be that these tireless, stripey-faced omnivores are planting the seeds of many fruits all the time.

14 May

The phone rings. 'A badger is dead on the road,' a friend reports, 'about half a mile away.' I go to see, hoping it's not from our local setts, although my gut tells me it will be. It's a young sow, her head crushed. The little eye mistily staring, jaw twisted and dirt-stained tongue protruding. No visible blood, no ripped skin – badger skin is extremely tough – just the signature of a massive, obliterating force. I stand and look down at the muddied shell of an extinguished life. Huge timber lorries cruise these glen roads laden with logs. Their wheels are massive. Cars surge through at 65mph. Even a badger's sturdy skull has no chance. I put the carcass in a fertiliser bag and take it home to bury it.

This is the norm. Every day badgers die on roads right across the nation. A shadowy form at the roadside seen too late, or not seen at all, a dull thud and the vehicle zooms on. Luckily they are prolific breeders and even the estimated 50 per cent kill rate does no more than temporarily dent the overall population. It remains a stark fact of unnatural history that almost everybody in Britain has seen more badgers dead than alive.

I examine her eight teats. She hasn't given birth. A virginal sow probably no more than fourteen or fifteen months old, in the prime of life, approaching her reproductive potential. If she had raised cubs the teats would have extended and developed mammae would be evident beneath them. The scent gland under her tail has stained the surrounding fur a nicotine ginger. It is pungent and seems excessively moist, as though in an involuntary last reflex she has ejected her entire reservoir of musk – a sort of olfactory last gasp. She was in excellent condition.

To have come through her first Highland winter so well she must have foraged energetically and profitably in the autumn, feasting on fruits, nuts and berries, grubs and earthworms before the cold set in. In recent years milder winters have been generous to badgers, resulting in a population expansion right across Britain. The same is true of the Highlands. Forty years ago, in an era of heavy snows and severe frosts I remember so well, Highland badgers struggled. Colonies were small, litters held to one or two cubs at most, often no cubs at all. Many badgers died underground, unable to forage in frozen ground and with insufficient fat to stave off the cold. The climate has changed – as for all of us – and they have made the benefits their own.

We have two local breeding colonies: one half a mile to the east, the other a mile to the west of us beside a disused quarry. We monitor both, these days with stealth cams. This sow almost certainly belongs to the east, close to where she died. I tip her into the hole and close it over.

28 May: 8.30 p.m.

The sun slips low and metallic beneath a thin gauze of reddening cloud. Still half an hour before we lose it behind the hills. Seventeen hours of daylight, stretching by four minutes every day. It's what happens here, north of the 57th parallel. I have come to the old quarry to see if I can watch the west colony badgers emerge, an act of reparation after the sow's death, almost as if I need to reassure myself that they are still there. I settle against a friendly tree trunk. Above me a cock chaffinch is shrilling out his claim.

In front, halfway up a steep bank twenty yards away, a truckload of sandy soil looks intentionally tipped, but no truck has been anywhere near this old woodland on the south-facing side of the valley. High above, a raven levers across the glen, then another, heading for the high moor. Their rough cronks collapse into the wood, stilling the chaffinch in mid-stream. Only the robin takes no notice. On the far side of the valley a tractor growls to a halt. It jars into silence and a dog barks excitedly. The tractor door slams with a short, sharp clop. On the croft pasture down by the river a cow coughs repeatedly as though it has something stuck in its throat. Then it stops.

Tree shadows reach and fatten like storm clouds. In three-dimensional ovals gnats dance to a tune only they can hear. Air thickens. Somewhere off to my left the robin tinkles its evening psalm and the cock chaffinch above me in the still-sunlit heights of the birch, and invisible behind a veil of new leaf, is silent at last.

In the grass at my feet a little brown moth flutters forlornly from stem to stem, seeming to achieve nothing. For a few more minutes the sun lances fiercely onto the birch stems, the silver-white bark fired to polished copper, leaves barred with luminous gold. Then it's gone, dipped below the burning rim of the hills to the west. As the glow of sunset fades a sullen twilight ghosts into the wood as if someone has draped a thin celestial veil across the sky; Highlanders call it 'the gloaming'. The robin's sad threnody to the dying day finally falters and stalls in mid-phrase as if he suddenly realises he's gone on too long.

Dusk gathers and pools. As shadows merge and ebb imperceptibly away, time stalls. Nearly an hour since the sun went down, the fiery rim is now only embers. I wait, forcing myself to keep still, a naturalist's discipline long practised. Above the heap of sandy soil is a badger sett, the entrance a circle of darkness. This colony is ancient; badgers have been recorded here for more than a hundred years, probably much longer. There is no wind, only gnats and midges. The western sky remains vivid, and the strip of horizon I can see through the trees is still gleaming as though lit with footlights. The angle of the sun's decline is so obtuse at this latitude that the gloaming stretches far longer than further south, a luminescent no-man's land between day and what passes for night in the Highland spring, perfect for watching badgers.

I fight off the urge to look at my watch. I can still see the earth mound, sandy pale against the looming black of broom and gorse bushes behind. My eyes and ears taunt me, playing tricks. The rustling of a shrew or mouse a few feet away make me think a badger is approaching through the undergrowth. Then I fancy I see movement, which dissolves to nothing. I try closing my eyes and drowsiness immediately closes in. I force them open again and focus on two pipistrelle bats robotically hawking back and forward above my head as though mocking me. One passes so close that I can hear the velvet flicker of its membranous wings.

There, without any signal or sound, suddenly a black-and-white striped face is filling the coal-black disc of the sett entrance. The black nose is inscribing Ms as it sweeps the evening air, a long and diligent scan for molecules of dread – of humans. After fully two minutes the face moves forward onto the earth mound so that I can see the whole animal, the grey body hunched and sturdy as a bulldog. It turns sideways and begins to scratch.

It scratches and scratches. First one hind leg, then the other, the black claws harrowing the long, coarse fur. Sides, shoulders and behind the ears, then rolling onto its side and scouring its belly, flanks and chest with the long, curved, combing front claws, twice the length of the rear ones, claws fit for a bear, specially adapted for digging. Badgers are plagued by fleas, and not much room to scratch properly underground. This is an emergence ritual: first test the air – take several minutes until safety is assured; move a bit further out onto the top of the mound – then scratch. Scratch all over and scratch some more. Keep scratching.

A second badger emerges, a sleeker, smaller animal. It sits, blocking the sett entrance. It looks around and also begins to scratch, but with slightly less vigour. I raise binoculars slowly and scrutinise every movement. Badger eyesight is poor and at this range there is no fear of slow movement being spotted. I pull focus and try to memorise the facial patterns and any scarring to head, neck or ears. Badgers habitually fight and often bear permanent scars on neck and rump; ears get ripped. I notice the first animal has a pale patch on its left flank – not an obvious scar, but a salmon-coloured stain, pinkish and the size of a book. I can't work out what it is. An old wound? A sore? I wonder if it's permanent or whether it has just been lying in damp sand.

I have scattered a few peanuts on the grass in front of me. I know they will scent them almost immediately. Badgers have a weakness for peanuts. The first animal has finished scratching; it stares at me intently although I am certain it can't see my shape leaning against a stout birch trunk fully forty feet away – eyesight too foggy for that – but so piercing is its stripy glare that I wonder if it can 'sense' my presence. Perhaps a molecule of my tainted breath has escaped up the bank and unnerved it. Or perhaps it is just the scent of peanuts it has detected. She – I feel sure it is a sow – ups and moves purposefully down the earth mound heading straight for me along a well-used path.

I am watching through binoculars; the image reminds me of a woodlouse, short legs invisible beneath the skirt of a shaggy coat. She seems to float weightlessly down the path, something stately about her gait, an importance parting the grass and cleaving the static air like a ship slicing through waves. She comes on; the distance is halved. She finds the first peanuts and I can hear the wet salivation of her chomping jaw. At fifteen feet she's too close for binoculars. I can lower them without fear of her seeing me as long as I make no sound.

The second badger is still sitting on the mound and shows no sign of coming down. It sits and watches for a few more minutes and then turns abruptly right and disappears, off on some foraging mission of its own. That makes me think the sow in front of me is a dominant female, not keen to share the peanuts and not to be messed with.

Alone with a wild badger a few feet away. I have enticed it to me with peanuts, unnatural delicacies planted like so much bait. Oblivious to my treachery and noisily guzzling, she is picking and masticating each nut in turn. There is no wind, although I feel sure she will soon cross my scent. She moves closer. I feel as though I could reach out and stroke her, but that would be as unwise as it is fanciful. The striped face is vivid with a purposeful glow.

If evolution has designed and equipped her with a warning flag, it couldn't be clearer – 'Keep your distance, I bite!' Even in soft gloaming light it gleams with a heraldic defiance. The white shines bright against the black and the little eyes are lost in the depth of the ebony blackness. Her head lifts sharply. She stops eating, jaw still and face fired with inspection, staring at me. I freeze. We are so close. Thirty seconds . . . forty . . . more . . . of deep, piercing inquisition. It seems inevitable that she will spook and flee. But no, whatever alerted her passes the hot testing of her sensory scanners and she resumes eating as though nothing had troubled her. The curious salmon patch on her flank is clearly visible.

I watch closely. I don't believe she can see the peanuts spread around in the grass, she is sniffing them out, one by one. She is nosing through the tufts of grass, swinging from left to right, panning, scanning, ever closer to me. Then she turns back as if she thinks she has missed one or two, but finds nothing. It's single-minded, this systematic search, occupying her whole attention for as long as it lasts. It's the tightly focused and relaxed behaviour of an animal that knows it has no predators other than man and she has satisfied herself; she thinks she has adequately checked out the human threat. I smile inwardly – I am undetected.

Deer don't feed like this; their long memory bank still tells them there are wolves out there. Ears constantly rotate and heads would jerk up wide-eyed every few seconds to test the wind and scan 300 degrees or more. I find myself wondering what the relationship between badgers and wolves was before we exterminated them in the eighteenth century. In northern Scandinavia wolves give the badgers' cousins, wolverines, a wide berth. I make a mental note to check it out. This badger is relaxed; she comes on. Only when she is almost at my feet does she turn away, still unperturbed. I want to get to know this badger.

When, a little while later, she does go, she ambles off in an unhurried, nonchalant but definite departure as if she has said to herself 'That's enough, now I need to attend to my foraging rounds.' I wait for a further twenty minutes to see if any others appear, then quietly sneak out of the wood. I don't need a torch. No wonder Highlanders have their own word for twilight; the horizon is still glowing and my eyes are well attuned, but a pallid darkness is slowly drifting in and a little wind whispers through the birches. High above me dark clouds are rolling by without a shrug, suggesting rain later on. Definitely time to go home. I pick my way carefully back to the track.

6 June: 9.00 p.m.

I'm back again. That salmon-coloured patch on her left flank troubled me. I hadn't seen a mark like that before on any badger, alive or dead. She had seemed fit and had behaved normally. I wanted to see it again, to grab a closer look. The hair on a badger's side is dense. It consists of soft, short underfur and long charcoal guard hairs with black and white tips merging to the grey of a foggy dawn. I hoped it was just discolouration, rather than the result of a wound or scarring beneath.

Full moon. I went with a strong torch with a red filter. Badgers are said not to be able to see infrared light. It was cool. It had been a clear spring day with the long, cheering sunshine of childhood memory, so strong that even glancing up into the tropical blue as bright as shining metal, my eyes could not take the glare.

Into that gladsome warmth daisies and primroses had beamed their broadest smiles. Bumblebees emerged from nowhere, as if summoned. Queen wasps awoke from hibernation under the slates of the roof and mooched hesitantly out to search for somewhere to start a nest. Soon their delicate papier-mâché lanterns would be dangling from the stables' rafters, an annual miracle of creativity that never fails to astonish me. The whole afternoon had hummed with the sprung energy of new life, new leaf, new colours, new scents, new opportunities all fuelled by the urgent current of gene-imprinted instincts, every one of them unimaginable aeons old.

Days like that fade to their close with a long, suspended sigh reminiscent of boyhood fishing trips I prayed would never end. I dawdled up the path into the wood. The spring night hovered, smelling of new grass, air rich with fertility. The sun's hot fingers had powered into the leaf litter and the clumps of winter-killed grasses, seeming to prise them apart. I trod warily to avoid the bulging spears of bluebells desperate to grab the light before new leaves closed off the wood-land canopy.

Wind carefully checked, I took my place against another birch. I needn't have bothered with the torch, although if I stayed long I could need it for leaving the wood later on. A few sprinkled peanuts and a wait. Not long – less than half an hour, still in good daylight, she arrived. 'Well, hello Dolly!' fizzed unbidden from my subconscious. It stuck. I never thought of or referred to her as anything but Dolly throughout the dozens of times I would see her over the next few years.

Badger-watching can become an obsession. Once you get to know the animals, and they become 'your' badgers, it is hard to resist going to see how they are faring. The naturalist, author and illustrator Eileen Soper, herself a life-long badger addict, once famously observed: 'Badger watchers are not entirely human.' Perhaps it's better not to be human, or at least to try. In Eileen's case habituation with her local badgers was so complete that cubs would take peanuts from her hands while others leap-frogged over her legs and set scent on her shoes. The celebrated nature photographer Laurie Campbell leaves his studio door open to the garden while he is at his desk in the evenings and the badgers come in and take food at his feet. I have never tried to habituate our badgers to that degree, but one of Dolly's offspring we named 'Stripey' became so familiar and relaxed that he would emerge in full daylight and calmly walk past us to clean up spilt grain under the bird feeders in the garden.

To watch successfully and have any chance of seeing relaxed badgers, you must sit downwind. Whether there is a detectable breeze when you arrive or not, your stance needs to be positioned where, if a wind does get up, it is in your face, blowing from where you expect the badgers to emerge, onto you – not the other way around. Even before scratching, an emerging badger will test the air for several minutes. After a long day of securely sleeping underground, the animal needs to satisfy itself that there is no danger out there.

If people or dogs have been near the sett, or have disturbed the land around it in any way, badgers will turn and head back underground, not emerging again until much later, or sometimes not at all that night. On the rare occasions that I have removed gorse or a broom bush to get a better view of a sett – a bit of discreet gardening – I then steer well clear for several days so that my scent has entirely evaporated and the badgers have checked out and accepted a slightly altered landscape.

Badger-watching on windy days is a gamble; better to wait for a windless evening. Winds swirl and eddy, especially in a wood where the air has to weave a path through trees. It makes animals twitchy. Prey species, such as deer, hares and rabbits, all rely on wind to bring them scent of danger. When that wind is swirling unpredictably, anxiety kicks in and they become jumpy. When a strong breeze is blowing through I've often witnessed red deer casting about, heads held high, ears rotating, nostrils straining, and then bunching together for the safety of numbers.

I didn't see Dolly emerge from her sett. Perhaps she was already out, scratching out of sight behind the bank of broom and gorse. She appeared soundlessly nosing across the woodland grass. She came very close. I could examine the salmon patch from only a few feet away. I shone the red-filtered torch onto her flank. In every respect she appeared to be a healthy, dominant sow. The unusual coloration seemed only to be an aberration, a freak of pigment rather than anything more sinister, valuable as a means of instant identification, even at a distance.

You sit and you wait and watch. Light drifts away, eyes adjust, pupils yawn to full wide, gnats dance until you can no longer see them. Your whole attention is harpooned by the animal in front of you. She becomes a part of you, an involuntary attachment hauling you ever further in. You lose control. You know you should be in charge but somehow you're not – you've lost it. She has gained the upper hand and now she holds all the cards. Her wildness is holding you captive, hanging on her every move. You know in your bones that if you let your attention slip, one slight movement and she will be gone. The spell will shatter like a glass dropped on a stone-paved floor and in an instant you will be sitting there feeling cold and foolish and alone.

Dolly stayed for twenty minutes. It's not so much a skill as an existential privilege to share that short time with a truly wild animal; an animal not tricked, but simply unaware that an arch predator is so close, so intimately pinned to its presence and its every move. Perhaps this is what it's like to be a tiger crouching in the reeds while marsh deer browse ever closer, or a weasel, as taut as a mainspring, eyeing up an unwary vole? My imagination is whirring like a windmill. Perhaps that's what I'm doing here: am I mobilising ancient predatory genes that have made humans stealthy killers for millions of years? Is my collective consciousness overriding the desire to kill and replacing it with a far more benign intent, but employing the same predatory techniques?

Hundreds of times over a long career working with wildlife, the behaviour of wild animals has made me ponder the unimaginable timescale of evolution. When did a badger become a badger? And what trick of natural selection tugged it off down that evolutionary avenue? When did humans start being predatory towards badgers? We know that mammals emerged some 50 million years ago, and mustelids were identifiable about 15 million years ago in the mid to late Miocene. The badger family, *Melinae*, emerging even later than that.

The idea of evolution is really very ancient indeed. The Greek philosopher Aristotle, for instance, some 2,350 years ago, one of the greatest intellectual thinkers of the western world, argued that all forms of nature constituted a series, a chain of being, which began with a very simple form and, striving for perfection, became more and more complex. A concept remarkably close to Darwinism, without the explosive idea of natural selection.

Many ancient hunting traditions believed that their quarry was in some physical and spiritual way related to them and that all life was connected – ergo the 'Great Spirit' of the North American Indians. Shamanism and the multiple other forms of magical or spiritual absolution were an important key to hunting success in many cultures. You worshipped the animal you were about to kill, requested its permission to give itself up to you and you dutifully honoured it by thanking it afterwards. Fetishes carved in ivory and bone celebrated the species of the hunt. But badgers are neither easy to kill nor particularly good to eat, so it seems unlikely that early hunters would have bothered with them until they began to view them as pests, consuming food of more immediate value to humans, such as cereal crops, game birds and their eggs and young. So the history of widespread badger persecution is probably much more recent, coming to a head in the Victorian sporting era when gamekeepers sought to exterminate everything that threatened their quarry.

Her head jerks up.

She is staring at me again. I can see her nostrils expanding and contracting as she strains every sense. I wonder what molecular cocktail she is analysing on those hypersensitive olfactory sensors. Can she tell that a heavily laden timber lorry chugged along the road below us half an hour ago? Can she detect the trail of polycyclic aromatic hydrocarbons that belched from its diesel exhaust? Does she know that a Border Collie was working the sheep in the river fields this afternoon and its crofter owner was smoking roll-ups? Has she sussed that a roe doe with twin fawns tiptoed through here early this morning? Or that the foresters working half a mile up the glen have had a brash fire and the smoke drifted down to us earlier on? Is it conceivable that Dolly knows that this afternoon Bella Macrae, a quarter of a mile away at the Craigdhu croft, was boiling elderflowers to make her delicious cordial? Or is she indifferent to any of the above, just on the *qui vive* for humans – any humans?

I read that badger noses are exceptional. Not only do they have hundreds of millions of olfactory receptors lining the mucus chambers in their noses, but the receptors also extend right up into the frontal sinus passages and sphenoid recesses. It seems they may have as many as half a billion receptors – half a billion! – that's way more than the hairs on its skin, all fizzing like champagne.

But senses don't work on their own. They collaborate. They are an evolutionary orchestration of everything coming together: scent, sight, hearing, wariness, pre-conditioned expectation, precautionary instinct, a deep memory bank, and very probably a finely honed animal intuition as well, all poised and set for action. Although badger ears are small – possibly as much to do with living underground as anything else – their hearing is also acute, as sharp as flint. Just what is happening here? Did Dolly hear my tummy rumble or did I breathe carelessly?

Her sooty little eyes won't help her unless I move, but hearing combined with those olfactory powers is formidable. An incalculable analysis is sifting through a thousand signals way beyond my comprehension. At this close range – now only six feet away – I can't hope to control molecules of my own pungent scent softly shedding into the evening air. I need a breeze to sweep them away from her, but there is none. It is a moment of shared foreboding, an uneasy stretched edginess for us both. She is galvanised by suspicion, unable to decide what to do. I'm as rigid as a rock. We share a common desperation: she desperate for clarity, for another confirmatory signal, and I desperately don't want to scare her off. We are locked onto each other and, for as long as it lasts, someone has thrown away the key.

I want to look at my watch, but I daren't. Time is elastic at moments like this, confusing me. Was it two minutes, five or seven? I have no idea. I started counting seconds in my head and then lost the place. Only her nostrils twitched. She was rooted, front feet firmly placed, head up, bright face stripes gleaming like a zebra crossing after rain. What is going on behind those stripes? What does a badger think? Or does it only react?

This is an animal that has no natural predators – certainly none in Britain. So all its fear is focused on us humans. To my knowledge no one has interfered with these setts during my time here, now more than forty years, so they have lived unmolested in these woods for many badger generations. Why, then, is she so wary? What has she to fear that makes her react so? What don't I know about badger history in the Highland glens? Were they persecuted for centuries before my time? We know they were sometimes trapped and snared to make sporrans from their stripy masks. Nowadays our local sporran-maker still uses badgers, but she sources them from roadkill. There's plenty of that, God knows. But since so few survive being hit, those remaining cannot know that human beings are driving the horrid killing machines. Instinct, I guess, takes millennia to develop and millennia to change.

Whatever finally decided her, I'll never know. Slowly she turned away and retraced her steps, unhurried, still snuffling about for missed peanuts in the grass, head obscured by her rounded shaggy back, until she merged back into the darkening thicket of broom and gorse. I sat motionless for a few more minutes. When I left the wood silence spread before me in a wave, the silence of fulfilment and the wave of triumph.

11 June: 2.17 a.m.

'What on earth was that?' Lucy sat bolt upright beside me. I struggled to the surface just as the dogs began to bark in the kitchen below. Then it came again, a second long, anguished scream ripping the night air apart, trailing terror through the short summer darkness like a bad dream. The window beside me is permanently wide open, summer or winter, a lifelong need to sleep with at least one small byte of my brain attuned to wildness.

Shorter and now closer, it came again primed with the sort of panicky distress you might expect from a six-year-old child stung by multiple wasps. A scream as much of fear as of pain, as edged as broken glass, shredded the cool silence of the night. This time I thought I knew its author. I jumped to the window just in time.

In the Highlands darkness never really happens in June. A lemon segment of moon shadowed the trees into looming martellos of blackness, air as still and dry as a cave. A lighted silver rim from northwest to north-east told both of the dying day gone and the slowly lifting dawn only an hour away. My eyes adjusted onto an expanse of lawn with just enough moonlight to render the grass to dull metal. I could see shapes, two shadowy shapes apparently connected. Two dark shapes lurching and dancing to a terrible yelling and screaming refrain. Lucy joined me at the window and whispered 'Whatever's happening?' So shrill were the screams I could barely hear her; they echoed against the house, surrounding us and sucking us into a vortex of knife-edged sound. A stranger might have thought that some dastardly atrocity was being committed right there in front of us, a slaughter of innocents. The cries were piteous, desperate, constant. 'It's two badgers,' I whispered, 'I can just make them out.'

It was a clash of badgers, a badger grapple, a badger dogfight, a scuffle turned fervent, angry and vicious. Perhaps it was an unintentional invasion of a big old boar's territory by a much younger, smaller male from a different clan. For years we've known that there are two local colonies a mile apart and that we sit midway between the two. Every year adolescent males get kicked out of both, evicted from their natal colony and forced to wander off to find a new home. Perhaps this young boar blundered in, lost, unaware and unwary. Did he cross the invisible boundary of scent markers, entering hostile territory, right into the jaws of the enemy – the wrong animal in the wrong place at the wrong moment? Perhaps he came round a bush and found himself snout to snout with a much bigger, stronger boar, who was having none of it.

As the youngster turned to flee, the big boar must have lunged into the attack. With sharply peaked canines he tore into the rump of the young male just above the tail. He bit deep and did not let go – badgers are famous for the power and grip of their jaws, muscles firmly anchored to the high sagittal crest along the top of the skull and a lower jaw permanently articulated in bone. The young boar struggled to break free and run, but couldn't. He screamed. He screamed that first gut-gripping, excoriating scream that had jerked us both from sleep. So in tandem they shuffled out from under the bushes, locked together, snout to rump, the younger animal still shrieking in distress, right onto the open lawn.

I could just make out the smaller animal continually spinning round to try to fight back, but he couldn't do that either. They circled and circled and the screams circled with them, fierce eddies of sound rising and falling against a ground of deeper, pitched rage through the teeth of the aggressor. It was a badger cacophony, a horrisonance of musteline ferocity, fear and pain. All we could see were shadowy shapes but the soundscape told us exactly what was happening. For fully twelve minutes this badger skirmish raged in increasingly anguished rings in front of us. Finally, bleeding and exhausted, the young badger tore free. In three seconds they had both vanished.

For a few more minutes we waited up to see if battle recommenced further away – but nothing. The silence of night poured in again, settling over the lawn like dew and an embarrassed moon slid behind ragged uplands of silver cloud. We returned to bed, breathless and shocked by the savagery of the brawl.

In the streaming sunlight of six in the morning I set out with the dogs to search for signs. The two Jack Russells quickly picked up scent and harried back and forth across the lawn creating frantic circles of their own. In places the turf was scarred by claws and I found dark clots of blood and a tuft of hair still attached to a skelp of hide the size of a bus ticket. The young boar had finally ripped free. Badger hide is extremely tough and I knew instinctively that if I found the wounded animal it would be in bad shape.

I phoned my field centre colleague, our Staff Naturalist, Alicia, who quickly joined me. 'It can't have gone far,' I told her. 'It must have been exhausted.' We rummaged through rampant rhododendron shrubbery, grubbed about under azaleas and into thickets of emergent nettles and brambles. The dogs locked onto the scent and barked hysterically. Finally, after ten minutes of searching the terriers homed in on a pile of corrugated-iron roof sheets, old and rusting, stacked against the stables wall. Alicia ran to get a cage trap.

Slowly and carefully we lifted the sheets away. There he was – there was the little boar badger, a cowering and trembling grey mop, his head invisible, jammed into the darkest corner he could find. He had been seriously beaten up, mugged badger-style. His wounds were angry and raw, the fur matted with blood. He had been savaged around the neck as well as his rump and one ear was badly torn. His coat was bloodstained and dishevelled – he was a mess. He didn't attempt to resist us; coaxing him into the cage was easy.

This is Fraser country, a Highland clan which, down the centuries, had fought many bloody battles of its own. *'Ifrinn an Diabhuil! A Dhia, thoir cobhair!'* ('Devil's hell! God assist us!') was their Gaelic war cry. So we named him Fraser. For two weeks we kept that little badger in a stable. We treated his wounds and he spent his days invisible beneath a deep pile of straw, only emerging at night to take the high-protein and carbo-hydrate foods we put out. He was meticulously clean, defecating purposefully in the farthest corner, always in the same place. He recovered quickly.

While Fraser was convalescing we constructed an artificial badger sett on a bank in the woods where we had never seen any signs of badger activity, hoping it was in no-man's land, well away from regularly patrolled territories. Ten feet into the side of a bank with three wooden, straw-filled box chambers four feet below the surface connected with twelve-inch plastic drainpipe tunnels leading downhill to two entrances fifteen yards apart.

As soon as Fraser was well enough we blocked off one entrance with a boulder and released him into the other, then blocked that too. Every day we put fresh food behind the boulder. Every day it disappeared. After a week we failed to replace the boulder and continued to place food in the open entrance. Every morning the food was gone. We removed the boulder over the second entrance and positioned stealth cams to record what was happening.

Badger aficionados had told us we had little hope of keeping Fraser on-site. Artificial setts often take years to be accepted and occupied by wild badgers. They were mistaken. Fraser stayed. Every dusk he forayed out and he was snugly home in his straw bed well before dawn. We continued to give him food and every week Alicia also put out fresh, sweet-smelling hay. Badgers love to spring clean their setts regularly. On camera we caught him kicking out the smelly straw and then, in a comical backward shuffling gait with a bundle of fresh hay tucked neatly under his chin, he reversed into the entrance and dragged it underground. He had found a safe home and made it his own.

16 April: a year later

On a spring morning bright with sun and birdsong, Alicia and I went to check the camera. We knew that Fraser was still in residence, but we hoped that, with maturity, he might have attracted a mate. To our great amusement we found that he had certainly got company, but not what we expected. A fox had moved in and occupied one of the other chambers, Fraser coming and going through one entrance, and the fox through the other. It is hard to imagine that they were friends, but it seemed clear that they were tolerating each other. That fox – we think a young dog fox – continued to live there throughout the spring.

24 July: that year

One morning Alicia came to tell me that, at last, Fraser had badger company – what Highlanders call 'a bidey-in', a mistress or an unattached courtesan. A slender sow badger was caught on camera hovering around the entrance before sneaking inside. She emerged again, scratched vigorously in the full sunlight of early morning, and then disappeared back underground. A few minutes later Fraser came trotting back from his nocturnal foraging and disappeared down the same entrance. We began to believe that we might one day get cubs from our artificial sett.

22 April: a year later

Yes, we did get cubs, but not from Fraser, nor from the artificial sett. It was Dolly who brought them to the quarry site. I had built a wooden hide, a substantial shed with fixed windows and padded seats so that we could watch in comfort and without the summer hell of midges devouring every inch of exposed flesh – an entirely different experience from leaning against a tree.

In the hide we could talk quietly, enjoy a flask of tea or even a glass of wine. We could stretch and yawn without fear of our scent floating off to the badgers. It meant that six or seven people could sit together and enjoy wildlife at close range, something it would be impossible to do outside. It's hard enough to keep still when you're on your own, but on occasions when I have tried with just one companion, we have been quickly detected. The hide was an instant success. Foxes, pine martens, brown hares and roe deer also came close. Hedgehogs tiptoed through and tawny and barn owls regularly came and perched nearby.

Three cubs appeared that evening. They must have been born in early February and only recently appeared above ground. Small, fluffy, silver-pewter, rumbustious and not a flicker of caution between them. They came rollicking in, barging, rolling, leap-frogging, bouncing, chasing each other in circles without a care in the world. Dolly arrived sedately a minute later with an obvious air of sufferance. They pounced on her, headbutted her, clambered over her, all primed with the irrepressible and carefree exuberance of youth. Over the next few weeks we were able to watch them grow and learn. The innate sobriety of adulthood gradually supplanted their playfulness. Beatrix Potter's depiction of Tommy Brock can only have emerged from a remarkably intimate understanding of natural history, a knowledge that must have been gleaned by first-hand experience in the field. She knew that adult badgers really are grumpy and short-tempered.

A few years ago a neighbouring farmer set a cage trap for a fox he believed was attacking his lambs. In the night it rained very heavily, a proper downpour which lasted for two days, causing the corner of the field to flood where the trap was set. I happened to be passing and noticed a movement in the trap. A large boar badger was up to its chin in water, and had been for many hours. Hardly surprisingly, he was extremely cross, perpetually snapping and growling at the steel mesh and at me. I carried the trap to dry ground before letting him out. As he shuffled off he snarled at me angrily, shook like a wet dog, and trundled away across the field, never looking back. It's a mistake to expect wildlife to say 'thank you'.

I would see Dolly's cubs many times over the ensuing weeks. I watched them grow and for the four more years she continued to bring her cubs to the hide site, I was able to witness the gradual, week-by-week disso-lution of their joyful cub exuberance, replaced by a far more staid and reserved foraging behaviour. A progression from fun to feeding, as they were gradually weaned.

Dolly is long dead. I had watched her for six consecutive years from that first emergence until one spring she failed to appear. She must have been eight or nine years old – a goodly span for a wild animal – and God only knows how many close calls she'd had with traffic on the glen road. As she aged her ability to accumulate sufficient brown fat each autumn to see her through the winter will have gradually declined. I suspect she died in her sleep deep underground in the grip of one of those excoriating frost plunges we get in most Highland winters, often lasting day and night for two or three weeks, when freezing air from starry nights floods down the mountains and pools in the glens. Frosts that freeze the river and layer the loch with eight inches of ice.

She had become a wise old badger and had more than done her work, raising perhaps as many as fifteen or twenty cubs, many of which will inevitably have pre-deceased her. But I am sure Dolly's genes are still with us, manifest in the many generations of badgers in our two colonies, wild animals with whom we are privileged to share our time and our space. Perhaps one day another badger with a salmon-pink patch on its flank will appear at the hide.

4

Pine marten

There is blood on the snow
and a trickle of rowan berry juice

on his bib where the pine marten
stands for a moment like a man.

'The Pine Marten',
David Wheatley (1970–)

Family: *Mustelidae*, the weasel family of the order
Carnivora. Subfamily *Guloninae*. Genus *Martes*, the
European pine marten. There are seven species of
martens worldwide, all in the genus *Martes*. DNA
studies reveal that martens emerged in the Miocene,
some seven million years ago. The British pine marten
is *Martes martes* from the Latin. Medieval English

mearth or *martryn*; Anglo-French *martrine* and *martre*; Old Norse *mörðr*; Finnish *nokia*. The old Scottish name 'martin-cat' was commonly used and the name 'pine' marten is somewhat misleading. While they are certainly at home in pinewoods, they occur across a broad spectrum of upland and lowland woodland and scrub habitats. They once occurred throughout Britain but were persecuted to a last stand in remote areas of the Highlands before staging a late twentieth-century recovery following widespread close-planted afforestation, which accidentally provided sanctuary. They are widely distributed throughout Europe.

The pine marten is a slender, extremely agile, chocolate-furred, cat-like mammal, adults weighing 1–2kg and measuring 51–54cm with a long flowing tail and the defining feature of an apricot-cream bib running from under the chin down between the forefeet. The bib is often spotted with one or two chocolate spots or streaks. Males are usually slightly larger and stockier than females. Fur is soft and silky, not unlike that of its mustelid cousin the mink, and it has historically been trapped for its fur throughout its range.

Males are dogs, females are bitches and the young are called kits. They have short, rounded, pale-rimmed ears and bright, dark eyes. Eyesight is excellent. Hearing is also acute and their sense of smell is well developed. They are acrobatic climbers with sharp, semi-retractable claws enabling them to climb trees nearly as well as squirrels. They are principally nocturnal, emerging at dusk and spending the dark hours foraging over a wide range, although they can often be seen out and about in broad daylight.

Pine marten habitat is typically in native woodland, scrub and rocky places from sea level to high in mountains, although they commonly appear in gardens, farms and around buildings and, if not persecuted, habituate to human presence very readily. In common with other mustelid species, they are very alert and inquisitive.

Martens are territorial and will fight fiercely and vocally to evict marauding martens from neighbouring territories with high-pitched screams and cat-like wailing. The kits screech perpetually when they first emerge from the den. They live individually or in small family groups while raising kits. Dens can be in hollow trees, rock crevices, in roofs and under floors of old buildings, even in old crows' nests. In common with all mustelids they have a scent gland beneath the tail and regularly and frequently bob down to set scent around their territory, the size of which is determined by food supply and can vary from a few hectares to huge forested areas.

Pine martens have a carnivorous dentition with well-defined canines and strong jaws, but enjoy a very varied omnivorous diet of cereals, fruits and nuts, as well as beetles, grubs and other insects. They are voracious predators, pouncing on their prey such as rats, voles and mice, squirrels, nesting birds and their eggs, young rabbits and amphibians, and will also take carrion in severe winters. If a marten gets into a hen house, it is capable of killing every bird. On bird tables they will consume almost anything from apple pie to bread and jam, honey or curried rice; peanuts are a great favourite. They remain very active throughout the winter and do not hibernate.

As with badgers, pine martens also possess delayed implantation, the genetic adaptation which allows the fertilised ova to be retained in the fallopian tubes until the marten bitch is in sufficiently good condition for the blastocysts to embed in the uterine wall and begin developing. This enables mating to take place in any month of the year, but typically in early summer with birth generally in January after a gestation period of seven weeks. Two to four kits are born blind and pink and stay in the den for around six weeks. When they emerge they are unsteady on their feet and entirely dependent upon their mother, often staying with her for up to six months. Kits can be very playful, with much scampering and tumbling over each other.

Today they have virtually no predators other than humans, although in historical times unwary kits may have been taken by wolves, lynxes, brown bears, wolverines, foxes, badgers and even eagles; nowadays the greatest threats are direct human persecution and roadkill.

Despite being a protected species, they have been widely persecuted by gamekeepers because they kill pheasants and partridges. Trapping and shooting have been the principal means, while illegal poisoning is also often practised. In the eighteenth and nineteenth centuries pine martens were mercilessly trapped for their fur as well as being outlawed as vermin on shooting estates, a practice that continued well into the twentieth century, driving them to the very edge of extinction. Their last stand was the remote glens of the Highlands of Scotland.

When, after the Second World War, the expansion of commercial forestry proliferated throughout the Highlands, ground preparation by deep ploughing broke through peaty soils and turned up furrows of nutrient-rich soils, which produced flushes of succulent weeds. This new food bonanza resulted in the dramatic expansion of voles and wood mice, in turn benefiting birds such as kestrels, hen harriers, barn owls and short-eared owls, as well as pine martens and wildcats. But as soon as the plantation trees established and the conifer canopy shut out the sunlight, the weeds and other field-layer vegetation disappeared and the voles and mice with them. The birds could no longer hunt and drifted away, but so widespread was the proliferation of forestry up every glen and across vast moors that pine martens were able to stage a comeback, continuing to exploit the dark forests as sanctuary. Now they are present in nearly every county in Scotland and also returning, partly by human reintroduction, to parts of England and Wales.

*   *   *

When I moved to live permanently in the Highlands in the late 1960s, the pine marten was an extremely rare mammal, one of Britain's rarest. I didn't hold out much hope of seeing one. When I did, I was spellbound.

Just a memory now, long ago, but still distinct. That first glimpse, that splashdown into realisation that such a creature still existed. The moment is sealed in for life. All these years later I could take you to exactly the spot up a remote glen in Sutherland. A moment so electric and unforgettable that even after all this time I could still tell you when – the year, the month, the day, even the hour. I'd like to boast that it spun me careening into instant awareness or understanding, or love or obsession, or something. But no, I'd be lying. It wasn't like that at all. It was a moment fogged with doubt, even disbelief – could it be . . .? Was that a . . .? Have I just seen . . . or am I fooling myself? Perhaps it was just a cat.

They were so rare back then, rarer than wildcats and as rare a sighting as an osprey – that now-cherished, cream-and-mocha fish-eagle gamekeepers had blasted to the brink of legend because they had the temerity to eat salmon and trout. That was the way of things in those days, a relentless, focused persecution of any raptor and any carnivore, a tyranny almost never challenged. And no one seemed to care. The martens had gone the same way. After centuries of being hunted by Highlanders for their glossy, mink-like fur, when

Victorian gentry re-fashioned the Scottish Highlands into an exclusive preserve for salmon, grouse and red deer, and gamekeepers were appointed as armed, over-zealous warlords to patrol and terrorise their vast upland estates, all predators were outlawed. Martens were trapped, poisoned, shot, hunted down with a singular and determined extirpation the only goal. That is why they became so rare. They were clinging on – just – by a claw and a whisker, and then only in the remotest mountain wilds.

So not so surprising that I couldn't believe my eyes, couldn't be sure that I had seen right. But it was. A furry dash across a twilit, single-track road somewhere up a treeless glen. And then nothing. It had vanished into an ocean of impenetrable gorse. I stopped the car, reversed back, waited, got out, ran forward, ran back, willed it to show again. Nothing. It had gone. But I knew in the marrow of my bones that I had glimpsed one of Britain's rarest mammals: *Martes martes*, a pine marten. That second-long sighting had hoicked it out of literary obscurity. No longer was it just a name on a page or a blurry photograph in a book, as remote and impersonal as a postcard in a gift shop. I had seen it in the flesh for myself. It had lived and moved and, for the blink of an eye, it had shared with me its agile, lissom, Bournville being.

That was then. Sometime back in the 1960s, when the Highlands was still a frontier of windswept, echoing emptiness. Many of its people had fled to the New

World or to cities for jobs, peeling away from the glens like blown leaves, leaving their pocket-handkerchief fields and tumbledown cottages to the sheep and the crows. And as always happens when you give nature half a chance, wildness had come clambering back, filling the vacant niches, shouldering in with what biologists call 'succession', a land- and a hill-scape heading back to the historical comfort of its own flowers and trees and shrubby scrubland, its own bugs and bees and secret arrangements underground. For centuries razed bare by half-starving Highlanders, the hills and glens were going wild again.

And then commercial foresters motored in, ring-fencing whole valleys and deep-ploughing the moors for Sitka spruce plantations. It became a craze, a plague, transmogrifying entire landscapes into the coniferous uniformity of cornfields. Yet, in the way that an ill wind blows nobody any good, those deep, unsightly furrows broke through the blanket peat and iron pans to provide a turnover of soils that allowed weeds and other nutritious grasses to burgeon and foster a brief population explosion of small mammals, a ready larder for pine martens and other predators, all of which would enjoy a short-lived glut of easy prey. As the young conifers grew to the thicket stage, they inadvertently provided an ever-darkening sanctuary for martens.

A decade slid past and that dark Sitka sanctuary yielded up an expanding population of martens, unseen, uncounted and unstoppable. The species was back from the brink, accidentally rescued by one of the most extensive land-use transformations to affect the uplands in centuries. Then as the conifer canopy closed, shutting out the sunlight, and the weeds and voles died away, the omnivorous martens were forced out into the wider man-occupied landscape for their food, but were still able to exploit the dark forests for breeding dens and sanctuary.

It was during those years that we established Aigas Field Centre in Strathglass. We didn't know it at the time, but that happy resurgence of pine martens was occurring right on our doorstep, in our own woods and other people's woods and plantations up and down the glen. Eventually, by the 1980s, a bitch marten had whelped in an abandoned shed beside the Aigas Lodge cottage – the first litter we found on home ground. Four enchanting kits entertained us that summer, popping up on bird feeders and delighting our guests. It would prove to be the beginning of a long and successful symbiosis, the martens enjoying our protection and our visitor numbers increasing year on year, as we became known as one of the few places in Scotland where martens could be reliably seen. We built a hide. To this day at Aigas many hundreds of people enjoy close encounters with this special mammal every year.

To suggest we've become blasé is wrong. One never tires of seeing pine martens. I have long since lost count of the individual animals I have known over the years, literally dozens, but it never palls. We see them at night from the floodlit hides, and we meet them face-to-face about the estate all the time. We never know where next. It could be in a stable or a barn, skipping along a balcony rail outside one of the guest lodges, checking out the hen run in the early morning, or nipping across a roof in broad daylight. We are never surprised; we share our lives with them and they bring a touch of flair and élan to our field centre days.

8 September

We had felled a wind-shattered spruce tree, Duncan and I, logged it for firewood and were stacking it at the back of our cabin on the loch shore, called the Illicit Still. A vital chore before winter sets in, fuelling the hut I built with my children when they were a wild, pre-teen gang always up for a fun holiday project. It is a fisherman's hideaway at the water's edge with bunk beds and a table, a paraffin lamp and a rusty oil-drum stove.

The children are middle-aged adults now with children of their own, but whenever they come home they head up to the Still, where I often go to write, embraced by a thousand happy memories. The lumberjack's stove is made from a forty-gallon steel drum and it is greedy – very greedy – but it belts out the heat. We need a good stack of dry spruce logs to heat the cabin on dark winter days when ice grips the loch and sleet on a lashing wind stabs wetly against the windows. Cutting, splitting and stacking logs is good, wholesome work. It renders up the tangy scent of sap, the white wood clean to the touch and the feel of scaly bark holding echoes of man's long past. A glow of satisfaction warms the heart.

The Illicit Still sits only a few yards from the forest where we had logged the tree. We were pleased with a job well done, chatting amiably, but as we emerged into the sunlight someone or something yelled at us. It was a thin, angry, cat-like yell. It cut us off. We stood in silence and gazed around. Then it came again, cross and fervid, and very close.

Like most mustelids, pine martens are inquisitive and bold. They are exquisite climbers, acrobatic as trapeze artists, and opportunists beyond redemption. They filch jam sandwiches from our bird tables – put out because it's such a delight to witness their agility and to see their shining, dark-chocolate fur with orange-cream bib, flowing tail, and their pert, intelligent little faces with eyes of bright jet. But this sandwich-eater is also a carnivore. On the rare occasions when Lucy has forgotten to close the hatch into the hen house when she shuts her chickens, guinea fowl and ducks safely away for the night, if a marten finds his way in the destruction is total – blood and gore in every direction, every bird dead or mortally mauled. In the chill light of dawn it's a level of carnage akin to a medieval battlefield.

This marten was not pleased by our intrusion. The shriek came again, and again. 'Clear off!' it seemed to say. 'Now!' At Aigas the martens see us going about our daily chores all the time and, in the absence of any threat, they have become accustomed to human activity. That familiarity has occasionally manifested itself in a level of marten contempt for us they make no attempt to hide.

He was twenty feet up a tree and within a foot or two of the twiggy nest of a woodpigeon. The parent bird had made a sharp exit, but two fat, half-grown squabs sat in the nest, dumb and defenceless. We had arrived at the very moment of his claim. His indignation surged to outrage as we stood our ground. The invective came as a drawn-out eruption of curse and damnation and the sort of abuse you might expect from a man who has come home and found his wife in bed with the dustman.

He snatched the first squab and ate it. We could hear the crunch of soft bone, the wet mouthing of hot flesh. Its limp red feet fell, one by one, to the needly floor. Then he grabbed the second, glared at us for a stretched moment of pure insolence with the squab clenched firmly in his teeth, and then slid up and away into the wood. Only his scratchy claws rattling on the pine bark gave him away; a smoke-curl of tail, a bark flake falling to the ground, an empty nest.

3 April: 5.37 a.m. – dawn

One of the field centre's young rangers, Charlotte, has told me she thinks a marten has a den in the wood behind the loch. For convenience and to avoid waking anyone else, I spent the night in the Illicit Still. I'm out at first hint of light, barely a streaky glimmer, to see for myself.

Dens are not always easy to find. If they're in a building, the martens coming and going usually give themselves away, but if it's a natural den site in a forest or a birchwood – perhaps in a hollow tree, among rocks or beneath roots – it is much harder to find. Pregnant bitch martens are secretive; they hole up some days before the birth. The dog marten plays no part. He has done his work long since and never goes near a den. Once the bitch is in, curled in a dry, cosy hollow, and hunkered down, fast asleep and waiting her time, you could walk past a den a hundred times and never know it was there.

A good friend at Ardnamurchan on the west coast, Michael Macgregor, an excellent photographer and naturalist, found that a pine marten had chosen to den in the roof of his house – it so happened, directly above the marital bed. When the marten whelped and the pups became vocal, the noise at night became intolerable. Mike knew that mammals often will move their kits to a new den, so very reluctantly he decided to cut a hole in the ceiling and gently remove the four kits, which were still quite small and immobile.

Very carefully he took them downstairs in a basket and put them out on his terrace in the sunshine. Then he stood back to watch.

Just as he had hoped, the bitch marten quickly responded to their calls and gently removed them in her jaws, one by one, scuttling off out of sight before returning for the next one. Satisfied that she was taking them to a new den, he went back inside to repair the hole in his bedroom ceiling. To his astonishment he found that she had beaten him to it. The four kits were back safely snuggled into the soft insulation wool beside the hole, only a few inches away from their former nest. That marten bitch had very clear ideas about where was best for her young. Mike caved in, repaired the hole and left them alone. What it did for domestic harmony is not recorded.

⋆   ⋆   ⋆

I'm batting in the dark. I have no idea whether she has whelped or not. The only information I've been given is that she has been seen in roughly the same place two or three times over recent days. Charlotte said she looked heavily pregnant. Scant intelligence, but I'm hoping I might see her returning to her den from a night's foray. My plan is to try to follow her or at least get a bearing on her direction. I know it's a long shot.

It is cold. What light exists is pure and shadowless. There was and still is a sharp frost. The loch is iced over and a stabbing chill assaults me as I step outside. I duck back in for gloves and a neck comforter. The stove had kept me cosy all night on an old sofa bed only four feet from the fire.

Once, long ago, on a winter expedition to the island of Barentsøya, Svalbard, locked into the pack ice of the Barents Sea to look for polar bears emerging from their dens with cubs, Lucy and I spent the night in an old bear hunter's log cabin. Outside it was minus 38 degrees Celsius and a swirling blizzard. All we could do was stoke up the stove and huddle down next to it in our sleeping bags. It's nothing like as cold as that here, but I applied the same tactics in the Still, wrapping myself in an old double duvet, far cosier than it had been in the Arctic. The loch ice glowed with moonlight from a cloudless sky, only a day away from full. I fancied I could hear the cold falling through a silver, shimmering darkness. I fell asleep dreaming of polar bears.

At dawn I stepped out into a rigid world of frost-printed patterns, nature's exquisite artwork. Everything is delicately gripped and fringed in rime. Leaves crackle underfoot. Dead bracken fronds are no longer ginger; each leaflet is rimmed in shimmering, translucent white. Pine needles form a frozen carpet as I stalk the edge of the loch, heading uphill. A sudden flow of cold air pierces all my defences, cutting deep. I shiver. Leaving the loch shore I wade through dead bracken thigh deep, praying I'm not making too much noise. I follow the burn, still gently running, the colour of brandy. Where the flow has splashed stones, frost crystals have flowered like delicate white fungi. I'm hoping the stream's gentle burble will mask my clumsy footsteps. I'm sure native hunters used rivers and streams in the same way.

I'm heading for where Charlotte said she last saw the marten, only another hundred yards to go, still uphill, weaving a slow path between gnarled old birches with deep bark fissures. There is a winding path running roughly parallel to the burn, but it's not human. I guess it's a deer trail, but probably also used by badgers, both of which are habitual, the red deer heading down to succulent grazing during the night and back again to the safety of the woods and the badger on his nightly foraging rounds. Where they have crossed the burn there is plenty of evidence. The twin arrowhead slots of hinds are everywhere, some very fresh, perhaps only a few hours old.

The ground levels out. There, on the rounded top of a boulder in a small clearing is a prominent marten scat, a twist of black excrement the size of a large caterpillar, intentionally and methodically deposited as a way-marker to other martens. It looks new, still shiny. Without disturbing it I can see mouse hair twisted in among scaly beetle elytra, as well as fragments of peanut, which tells me it has fed at one of our bird feeders or hides. Encouraging.

Despite the cold, I decide to sit, hunkering down. I breathe slowly and deeply, in through the nose and out through the mouth, part of the disciplined process of achieving stillness. My breath plumes and I wonder if molecules of me are slowly spreading through the trees. I muse about how far they go before they disperse, swallowed up in the fungally essence of the woods. I pull my neck comforter up to cover my mouth and nose and fold my arms across my chest, tucking in to trap as much heat as I can. Somewhere, down by the loch, the silence of dawn is saluted by the echoing hoot of a tawny owl, a last shout before going to roost. It goes unanswered.

Hunkering down is a discipline. It is a personal technique learnt from an elk hunter in northern Finland many years ago, later endorsed by Eric Ashby, a wonderful naturalist and filmmaker in the New Forest, who briefly took me under his wing when I was an enthusiastic student. He was later dubbed 'The Silent Watcher' for his endless patience when filming wild animals.

It is something you can only do on your own. The hunter, who was half-Sámi, half-Finn, explained that by far the best way to hunt was to let the elk come to him. He would sit motionless for several hours if need be, moving nothing but eyes in a Zen-like dedication to stillness – a denial of self, a meditation above and beyond meditation. 'You must expect nothing,' he added. 'What happens if you get an itch?' I asked. He laughed and told me that you had to teach yourself to let the itch pass through you and out the other side.

If you gave in to it and scratched, you burst the bubble, then very hard to recover true stillness and the awareness that stillness delivers. The itch would be the test of your resolve. As he told me this he was shaking his head and smiling. I sensed that he didn't really believe that I could do it, and that it was a state of practised being, and an awareness of 'self' out of reach to 'normal' folk, personal only to him and his Lapp hunter friends. It made me all the more determined to achieve it for myself.

I struggle to describe what is really happening here. It is tempting to speak of meditation, but it isn't that at all, and nothing to do with a trance. Most Buddhist-like meditation is about clearing your mind and shutting external awareness down. What I'm referring to is almost the reverse: achieving a hyper-awareness of surroundings and self as one, and I am the first to admit that I don't properly understand it or the laws that govern it. Nor does it happen immediately; you have to practise it, over and over again, as one would to achieve a full state of meditation. An inadequate analogy might be my laptop blanking itself to save battery power – not shutting down, everything is still fully functional inside – but temporarily on hold until something jogs it back into life. The only parallel I can find with Zen meditation is that of controlled breathing, essential for both disciplines.

Like the elk hunter, I think it is only achievable if you are fully aware of your habitat: an awareness attuned to its sounds and scents, the pressure of air on your skin, the feel of moisture or lack of it in your surroundings, the acceptance of ambient temperature, the 'sense' of rock or soil or woodland litter beneath you, the acknowledgement of branch and leaf movement by wind or breeze, the shifting light and shadows, the abandonment of body and the denial of time passing.

To expect a person from an urban habitat to understand what I'm talking about is a huge ask. I have often taken folk out into the wild woods and asked them to keep still while watching wildlife such as deer, only to find that they are incapable of not just achieving stillness, but also failing of understanding what stillness really is. I am reminded of what Annie Dillard (1945–) states in *Pilgrim at Tinker Creek* – 'It is astonishing how many people cannot, or will not, hold still. [. . .] at the creek I slow down, center down, empty [. . .] I retreat – not inside myself, but outside myself, so that I am a tissue of senses. Whatever I see is plenty, abundance. I am the skin of water the wind plays over; I am petal, feather, stone.'

You can't scratch; you can't sneeze or cough; you can't blow your nose or shift your position – you have to learn to control all those bodily actions we normally don't even think about. Nor should you go to sleep, although I confess I sometimes do, by mistake. Eyes need to stay fully open – along with the rise and fall of breathing, the only actual bodily movement allowed. And you have to give it time – open-ended time – for as long as it takes. Even keen young students of wild-life conservation look nonplussed when I tell them they may have to sit very still out in the woods for several hours. Guests in the comfort of our hides, warmly wrapped and seated on padded seats, look shocked when we tell them they must sit very still for up to three hours.

Hunkering down is a technique I've used many times and had some memorable experiences as a result. It is as though wild animals are hardwired to expect humans to move about and to be noisy – that if you are properly still and silent, you can't be human. After a cursory inspection they ignore you. I remember Eric Ashby telling me 'There isn't much you can do about your scent, but most wildlife requires more than a whiff of scent. They need sound or movement to confirm human danger.'

Frolicking badger cubs and pine marten kits have blundered into my feet, birds have landed on me, deer have often grazed right up to me, weasels and stoats have busied about around me, brown hares have stumbled into me, red squirrels and wood mice have scuttled over me, and just once, in our own Aigas pinewood, a wildcat stalked silently past three feet away, carrying a leveret in its jaws. I am sure it didn't know I was there.

*　　*　　*

An hour and a half drifts past like a mirage of time lost. A steely light has filtered in through the trees. Great tits are out and about in the canopy above me and a coal tit came down to check me out, then hopped about searching for spiders in the birch-bark fissures. Nothing else has moved. The cold has wicked up from the frozen moss I'm sitting on and chased all my body heat into its core. I'm so cold that it takes gritted teeth and all my will power to stand up and get my limbs to work. I teeter slowly back downhill. As an after-thought I go back and remove the scat from the boulder, keeping it in my gloved hand until I get to the loch before tossing it away. If it gets replaced I'll know my marten is out and about.

Sometimes I think that trying to be a naturalist is a test too far, altogether too demanding. 'To hell with bloody pine martens,' I mutter to myself. I want a hot bath and a mug of steaming tea, then a bowlful of porridge with cream – I can't think of anything else.

7 April

Another try, a second night in the Still. Same routine, supper at home with Lucy and a goodnight hug before slipping out into the cold. The Still stove quickly roared into life, spruce logs crackling like pistol fire in the flames. I pulled the sofa bed close before wrapping myself up in the old duvet, fully clothed. Sleep creeps in like a flooding tide when you're cosy and warm. Everything I needed for dawn laid out on the table and boots on a chair close to the stove. At least I can start out warm; there's nothing worse than forcing your feet into cold, damp boots. The last thing I remembered was wondering how long before the moon rose.

No need for an alarm clock. If my brain shuts down knowing that I want to be up before dawn, it seems to have an infallible clock of its own. At five o'clock I was wide awake. Just time to clean my teeth, boil the kettle and gulp down tea before heading out.

Thankfully not so cold. The moon is past full, waning gibbous and lurking behind chiffon rags of cloud. The still-frozen loch has the bright shine of a silver shield; the ground still crunchy from savage frosts past, but a velvet breeze from the west is lenient and brings promise with a first hint of spring in its wake. Light hangs in the east like a hesitant guest. A robin is doing its morning thing from a goat willow right outside the Still back door – robins never seem to miss a trick. This time I know where I'm headed and walk briskly up the burn.

When I arrived at the boulder there was no fresh scat. My spirit slumped. I had hoped for a new signal of marten presence, but just as I was wondering whether I was heading up a blind alley, a movement caught my eye.

We use the expression 'out of the corner of my eye' all the time, often without realising that the human eye is cunningly designed to capture movement at the extremities of normal sight. Of the two different photo-sensitive cells on our retinas, rods and cones, the cones work best in good light (photopic vision), they detect colour and give us sharp focus. Cones are exclusively concentrated in the centre of the eye, whereas the rod cells, which are concentrated at the sides, primarily detect monochrome and work better in low light (scotopic vision). Rods arranged vertically like railings pick up movement detected at 'the corners' of the eye. Just as movement is accentuated by a passing car seen through railings, similarly a movement passing over rods catches our attention even though we haven't seen what that movement was. To see it properly and to pull focus, we have to shift the object of movement to the cones.

That's what I did, slowly and carefully. And yes, it was a pine marten.

On the ground, martens have a particular 'lope' all their own. It is an efficient, easy, lolloping gait that looks, and probably is, relatively effortless. I can't liken it to anything else, certainly not a wolf's jogging lope, nothing like a prowling cat or the bouncy trot of a fox. You can't over-emphasise the coiled-spring impression a marten's lope contains; within that relaxed bounding action it possesses an invisible hair trigger for sudden explosive action. They can spring five feet into a tree and vanish vertically upwards so quickly that if you blink, you miss it. That is what I saw.

It was thirty yards away. My rods had logged it loping behind and through the birch trees. As I turned, my cones pulled focus on an adult marten springing onto a birch trunk and vanishing into its upper branches.

I knew straight away that it had seen me. It's hard to fool a marten.

So I was faced with a choice: to sit it out and hope that its innate curiosity would bring it down to check me out, or to give up and go home. At moments like that it's too easy and very tempting to capitulate, to shrug the shoulders and say 'Oh well, I tried.' But like the elk hunter, I elected to stay, to hunker down and hope. I settled beside the boulder, my back against the papery bark of a young birch, I pulled up my collar, tugged my hat down to shield my face and briefly closed my eyes to focus on not being there.

I don't know how long it took. After a while time dissolves into itself and ceases to be of any consequence. After an hour or two you lose your identity; you become rock, or tree, or fungus, winter-killed bracken or just plain earth until inevitable bodily sensitivities haul you back – cold, cramp, hunger, thirst, bladder . . . A cold sun lifted slowly over the loch and teased its way through the birches and pines, tiger-striping the needly floor and side-lighting the trunks of trees as if painted in gold leaf.

A centipede crawled over me and I tried to guess at the number of its legs; a tiny spider descended on its silken thread from the brim of my hat, so close that I couldn't pull focus. A flutter of tseeping coal tits busied around me just out of vision, and a large, shiny black beetle, a dor beetle, with a luxurious blue metallic iridescence to its carapace, ponderously six-legged it past me on the forest floor, in a straight line, as though it was on a mission. Dors are dung beetles and I wondered if it had detected a scent signal from the scat I had removed. In nature everything is connected; it is only us who cut ourselves off, shut ourselves out.

I didn't see the pine marten again and I would have had to write my field notes up as a failure, had it not been for a sudden and electric flash of woodland drama.

I had been aware of jays in the wood. Their frantic screeching calls echoed from the distance all morning, a raucous conversation edged with petulance. I had wondered if they were alarming at the presence of the marten, but the exchange of calls seemed to be too persistent, too much two-way bickering between birds, like some testy territorial dispute. And then it happened. One jay flew quite close, only a few yards away, its white rump flashing like a beacon, landing somewhere only just in my view, up in the branches of a pine. Perhaps I was wrong, perhaps after all the jays were very well aware of the marten presence – a potential predator of their eggs and young. Jays are wary birds, much persecuted, so I knew from its proximity that it had not detected me. Perhaps they would reveal the whereabouts of the den to me. Its screech filled the whole space around me, silencing all the other birds.

I turned my head slowly and looked up just in time to see a large dark delta-shape come plunging through sunlight. It flashed past me, twisting through and up into the tracery of spare winter canopy. A shriek sliced into the silence like a sabre. It was a shriek of pain and desperation, of shock and alarm, and of the inevitability of death.

A large female goshawk had snatched that jay from its perch. It was an ambush, a sudden unseen rush of bird, a rippling slash of wings and talons out of nowhere, a missile-fast snatch and grab, so devastating that the jay had no time to react, no chance of flight. Marigold eyes burning, the big hawk came fast from only a few feet above ground level, twisting and turning on thrusting, half-closed wings, flashed past me through the trees only a few feet away before veering sharply upward and snatching the jay from below. The long, curved talons of one foot, hypodermic-sharp, struck just once, grabbing the jay by its pinky-fawn breast. A second later the goshawk was gone, skimming away through the trees with its prize as the wretched jay screamed a long, last wailing cry.

Death in the talons of a goshawk comes slowly. The hawk would have flown to a convenient branch, perhaps one it had used many times, and with the jay pinned down in one foot its hooked beak, scalpel-sharp, would have ripped at the jay's neck, plucking feathers and skin to expose the naked flesh and the jugular vein. No amount of thrashing wings or screeching would have deterred the steady, practised execution of the jay.

I have seen goshawks many times, but I had never witnessed that twisting killer attack through the trees.

The power, the speed, the lethal accuracy of the snatch had taken my breath away. And then it was gone. I had to shake my head to be sure I hadn't dreamed it. My pulse was pounding. I sat on for another few minutes before silently slipping out of the wood.

11 April: 6.40 p.m.

With daylight slowly lengthening, I thought I would try the evening, when a pine marten bitch would normally emerge to feed. I was convinced I was on the right track, so I returned quickly to the boulder beside the burn path. Yes! A new scat. Fresh marten scat issues a sickly sweet scent. It was only a few hours old. Somewhere, I think uphill, probably not very far away, there is a den. All I can do is proceed very carefully and slowly, as quietly as possible, in the hope of coming across some other sign, some give-away mark or sound – as so often in natural history, really down to those two essentials: chance and perseverance.

A robin followed me, flicking from branch to branch, hoping I would disturb a bug or insect it could snatch. I find myself wondering how many hundreds of thousands – perhaps millions – of years ago robins learnt that trick, to dog the footsteps of humans for an easy meal. A red squirrel saw me coming and slipped scratchily away, first out of sight round the back of a pine tree, then up and away into the needly crown. I walked on knowing full well that it was watching me.

It reminds me (as if I needed reminding) that it is virtually impossible to move through a wood without being spotted and marked down by the wild creatures with which we share our space. It is our human burden, our perpetual millstone and the price we pay for our relentless domination of wildlife. I have always seen it as a personal obligation to do penance, to try to accommodate wildlife in every way I can.

The ground rose steeply with rocky outcrops beside the deer path I was still following. I sat for a few minutes to catch my breath. A hoodie crow flew overhead cawing abrasively. Hoodies are the perpetual spoilsports of the forest. They taunt you with their rasping cries, telling everything around that you are there. I cursed it silently and decided to sit tight in the hope that it would get bored and move on. I made myself comfortable against a smooth rock, stretched out my legs, closed my eyes and hunkered down.

Sleep was not part of the plan. When I awoke I felt a little foolish, wondering what I had missed. The wood was eerily silent, the low sun long gone behind the mountains and the light ebbing away like a tide. I realised I must have slept for an hour or more. I wish I could say that it doesn't happen very often, but in truth it does.

I believe it is incumbent upon non-fiction writers, especially nature writers, to say it as it is. Truth matters, otherwise we might as well make it all up. So, yes, I had slept soundly. I just find it very hard to stay awake when striving to keep completely still, especially if it's cold. As I have got older sleep seems to come more readily, sneaking in and involuntarily shutting me down so stealthily that I don't feel it coming and before you know it, I'm gone. In this instance, sleep was not such a bad outcome. It meant that just about everything in the wood that had been alerted by the hoodie crow forgot about me and shuffled off about its own business.

My legs were dead and rigid. I flexed my calf muscles and wiggled my frozen toes inside my boots to get some blood flowing again. Years of experience have taught me never to get up and stretch until I'm sure nothing is close by and watching. I could check out 180 degrees of vision without moving my head. Seemed all clear. Then, just as I was about to ease myself up, I heard the distant but unmistakable prattle of an alarming wren. I froze and listened intently. It was somewhere up ahead, some way off, chittering angrily. You can't mistake the call. It sounds cross and slightly hysterical in a thin, peevish, grating rattle. The tiny bird was flicking from tree to tree, out of sight, but never going far away and returning to roughly the same spot over and over again. Something had upset it.

Reading nature's sounds and signs has always excited me. Hundreds of times over the years of being an active field naturalist the alarm calls of birds or the gruff barks of deer or even the sound of snapping sticks or rustling vegetation have alerted me to moments of wildlife drama or just wonderful sightings and experiences I might otherwise have missed. Instinctively I knew that the wren was not alarming at me.

A dilemma. If I stood up and attempted to creep up to see what it was, I ran the risk of seeing off both the wren and whatever was troubling it. If I sat still I might miss it altogether. As a compromise I elected very slowly and carefully to shift my position so that I could scour up ahead through binoculars. It worked. A pine marten was just visible halfway up a Scots pine, perched on a large branch, busy scratching behind its neck with a rapid rear-foot action. I was pretty sure it had just emerged from a den, which must be very close by. It had not seen my movement.

After scratching for more than half a minute the marten slid elegantly down the pine, head first, to the ground. I could hear the claws of its hind legs, splayed out behind it, snagging against the bark as it eased effortlessly down the trunk onto the ground. In two bounds it had disappeared. I grabbed my chance, stood up and, carefully avoiding snapping any sticks underfoot, I quickly crept forward.

The wren saw me immediately and continued to alarm, now definitely at me. I ignored it on the assumption that the marten probably wouldn't register the difference. Five yards, ten yards, fifteen. Stop. Scour ahead. Nothing moving. Another five.

Stop again, scan left and right through binoculars – still nothing. The wren flew close and then off again to a safe position among the branches of a fallen birch beside the burn, still prattling loudly. I knew I was close, perhaps very close.

The marten had vanished. Daylight was vanishing fast. I didn't know what to do. Casting uncertainly around I couldn't see a hole in a tree; there didn't seem to be any root hollows or tunnels anywhere; nothing but a slowly darkening wood beside a softly burbling burn. Then it came.

Off to my right the ground fell sharply away into the bed of the stream. Large boulders lined the banks like a mini gorge, the result of angry winter spates when steep burns like this roar in a slow, perpetual thunder. From somewhere among those boulders came the clear bleating call of a pine marten kit. Once you've heard it you never forget it. It is vaguely cat-like but not as high-pitched, a monotone bray edged with pleading. Another and another – I guessed three separate voices calling for their mother, repeating over and over. And there she was, at the water's edge, not twenty feet away from me, calmly grooming her chocolate fur.

The three kits emerged and stumbled towards their mother. They looked as though they were new to the world – that this might have been their first foray out from the den in the rocks. The bitch marten hadn't seen me, was wholly unperturbed. The wren was still prattling from a safe distance. Slowly and nonchalantly she threaded her way between boulders, heading down-stream, only pausing to make sure the kits were following her. In less than a minute they were gone, although I could still hear the kits' bleats fading until they were lost under the babble of water. Only then did I move down to inspect the den.

22 May: 8.35 p.m.

Six field centre guests had gone to the Campbell hide, a comfortable, green-painted timber shed in the woods named after Laurie Campbell, Scotland's celebrated nature photographer who has so expertly led our Aigas photography courses for the last twenty years. The hide has LED floodlights throwing a pool of daylight onto a manufactured feeding station made to look as natural as possible, a site frequented most nights by badgers and martens. Wood mice attracted by the peanuts bring in tawny owls; roe deer tiptoe through and occasionally a fox, a brown hare or a hedgehog pop up to check it all out. The martens and badgers take no notice of the artificial light raised gently on a dimmer; our guests get a really good chance to see wildlife at very close quarters, through open windows, often only a few feet away.

Last night they were in for a surprise. An adult sow badger came shuffling in and began to gobble peanuts. To everyone's delight, a few minutes later a well-known pine marten bitch we have named 'One Spot' because of a dark chocolate spot on her bib, appeared with two wobbly kits, not long out of the den, probably only six or seven weeks old. Young marten kits call continuously to their mother in the early days of emerging from the natal den at about six weeks. It is an unmistakable, persistent scratchy cry, the purpose of which would seem to be for the mother marten to know exactly where her kits are as they venture through the woods.

One Spot nipped up the branches of a fallen tree we had carefully positioned for photographers in front of the hide windows and began to feed on honey and peanut butter one of the rangers had smeared in several suitable positions. The kits tried to follow her, but were still unsteady on their feet and their climbing skills distinctly unpractised. They clambered clumsily up, fell back, tried again and called out loudly. One Spot ignored them. She was busy licking honey from the highest branch. One kit then fell to the ground and wobbled off to a hollow stump where the duty ranger had placed some concealed peanuts. It climbed inside. The watching guests were utterly enchanted.

At that point the badger, which had so far ignored the martens, moved purposefully to the stump where it lunged at the marten kit, mustelid attacking mustelid, grabbing it by a back leg. The kit screamed and tried to scramble away, but the badger shook it angrily. One Spot leapt down from the branch and began to run off into the wood with the other kit. The badger let go and the bitten kit limped after its mother and sibling, clearly injured and wailing pitifully.

The guests in the hide were horrified. One lady burst into tears. Another wanted to throw something at the badger. The duty ranger, a very conscientious lad called Louis, did his best to calm them down. He told me later that while he didn't think the marten kit was badly hurt, the guests were clearly in shock. Our glossy television culture of nature viewing doesn't expect wildlife to be wild and susceptible to injury or death, certainly not while being watched from a hide, especially not if it is young and cuddly.

I have often seen tension and aggression displayed between badgers and martens, but usually, of course, the marten is far too quick and agile to be caught. They may be cousins, but there is no love lost between *Meles* and *Martes*. If they could I am sure badgers would kill martens all the time. In the event that a badger happened across an accessible den – among tree roots or a burrow – I am certain it would dig the young out and devour them all just as badgers do with nests of rabbits, mice and, of course, weasels.

Both One Spot's kits survived into adulthood, although the bitten kit did limp for several weeks.

5

Otter

Family *Mustelidae*, the weasel family of the order *Carnivora*. Subfamily *Lutrinae*. Genus *Lutra*, Eurasian otters. There are thought to be thirteen species of otters worldwide, divided into three separate genera. Only the genus *Lutra* occurs in Britain. The British otter is *Lutra lutra* from the Latin. Medieval English *ottr* or *otor*; Scottish colloquial *dratsie*; Scottish Gaelic *dòbrhan*; French *loutre*; Danish *odder*; Spanish *nutria*; German *Ottern*. It is thought that the word 'otter' stems from a proto-Indo-European word, *wódr*, from which we also get 'water'.

There is often confusion between the terms 'river otter' and 'sea otter'. Sea otters are an entirely separate New World species, which live exclusively in the Pacific Ocean, largely feeding on abalone shellfish.

British otters are as much at home in salt water (hence the confusion) as in rivers and lakes, on land and in the sea. They are very agile on land and often travel considerable distances away from water. Otters were seriously threatened by organo-chlorine pesticides and pollution of rivers in the 1960s and '70s and declined rapidly. As toxic pesticides were gradually banned and rivers cleaned up, otters staged a remarkable comeback. It is now thought that otters are present in every river in Britain. They are also widely distributed throughout Europe.

The otter is a thick-set, elongated, extremely agile, brown-to-grey furred, semi-aquatic mammal with webbed feet on short legs and a long thick tail known as a 'pole'. Adult males are much larger than females. Males weigh up to 10kg and females 7kg. Fur is soft and silky, with dense, waterproof underfur and long outer guard hairs which give the animal a ruffled look when wet. Otters have historically been hunted for sport, persecuted for eating fish and trapped for their fur throughout their range.

Males are dogs, females are bitches and the young are called cubs or pups. They have short, rounded ears and bright, dark eyes. Eyesight is excellent. Hearing is also acute and their sense of smell is well developed, although of no value underwater. They are well adapted for an aquatic life with webbed feet. It is thought that in common with seals, their long facial whiskers, vibrissae, can detect fish movement and are a hunting aid in murky or muddy waters. Otters are often nocturnal, but equally happy feeding by day, particularly in the sea where tides are a significant factor.

Otter habitat is typically in and around water, both salt and fresh. When otters disappeared from polluted rivers those otters living on the coast and on offshore islands, particularly in the north of Scotland, became the remnant populations from which recovery and re-colonisation could take place. Seeing a wild otter used to be a rare event, but in recent years otters habituated to human presence have regularly appeared on rivers, lakes and ponds in towns and cities throughout Britain and they not infrequently appear in gardens, farms and around buildings. Boat owners have often arrived at their mooring to find a sleeping otter curled up on a seat. Just like all the other mustelid species (except perhaps the badger), otters are very alert and inquisitive.

Cubs are born in a burrow nest called a holt. Holts can be in hollow trees, among roots or rock crevices, even under upturned boats or beneath the floors of old waterside buildings, and are often lined with dry grass. Two to four cubs stay with their mothers for many months and can be heard whistling with a thin, plaintive call when they first emerge from the holt. They live individually or in small family groups while raising kits. In common with all mustelids they have a scent gland beneath the tail and regularly and frequently bob down to set scent around their linear, almost always riparian or coastal territories, the size of which is determined by food supply and can vary from a few miles of river to whole large lakes.

Otters have a carnivorous dentition with well-defined canines and strong jaws, but enjoy a very varied diet of fish, crustaceans such as crabs and lobsters, amphibians like frogs and toads, but also small mammals, and ducks and ducklings and the eggs and chicks of other aquatic birds. In Scandinavia they are known to enter beaver lodges and prey on beaver kits. Gull and tern colonies are particularly susceptible to otter raids, when they consume both eggs and chicks. They are voracious predators, capable of causing havoc in a hen run and making off with domestic ducks and even geese. They will also take carrion in severe winters. They remain very active throughout the winter and do not hibernate.

As with badgers and pine martens, otters are also thought to possess delayed implantation. This enables mating to take place in any month of the year, but typically in early summer with birth generally in January or February after a gestation period of nine weeks. Two to four cubs are born blind and pink and stay in the holt for around nine weeks. When they emerge they have to learn to swim and are entirely dependent upon their mother, often staying with her for up to a year or eighteen months. Once they can swim, cubs can be very playful, with much diving and tumbling over each other.

Today they have virtually no predators other than humans, although in historical times unwary cubs may have been taken by wolves, lynxes, brown bears, wolverines and even eagles; nowadays the greatest threats are direct human persecution and roadkill.

Despite being a protected species, they have been widely persecuted by gamekeepers because they occasionally take pheasants and partridges. Trapping and shooting have been the principal means, while illegal poisoning is also often practised. In the eighteenth and nineteenth centuries otters were systematically trapped for their fur as well as being outlawed as vermin by anglers, a practice that continued well into the twentieth century.

Now, however, otters are back. They have re-colonised just about every suitable habitat in the UK; thanks to dedicated conservation work, significant works of literature and natural history programmes on television, they have entered the public psyche as a valued and important member of our mammalian fauna.

29 June

Bright sun at midday. Twenty-four degrees Celsius. An occasional flurry of breeze ruffles the water on the loch, then dies away. With barely audible swirls, trout rise lazily to the latest fly hatch, leaving expanding rings gently undulating on a glassy surface. A mallard duck has just hatched nine chicks – this late in the season it must be her second brood – and is weaving them slowly through the sedges at the margin of the marsh. They rush around her in tiny spurts of energy as they snatch flies from the surface. Just out of sight in the pinewood that skirts the southern shore, I can hear a group of primary school children laughing and shrieking.

I was working at my laptop in the shade of the Illicit Still, where I go to write. My phone pinged. A WhatsApp from one of our education team in charge of the school kids tells me they've spotted an otter swimming along the edge of the marsh, diving and rising again as it works its way through floating mats of *Potamogeton* weed.

Lottie is twenty-one and a student of Conservation Biology. She's an academic placement with us for the third year of her course at Exeter University and is busy researching our beaver demonstration project. She had been monitoring beaver activity when she saw me working in the sunshine and swung by after setting a camera trap near the beaver lodge. I pointed to the marsh at the far side of our little loch. 'Otter,' I mouthed. She joined me on the deck; through binoculars we could clearly see its glossy head and bow wave.

This is not unusual. Otters visit the eight-acre loch regularly all the year round. In winter I find their footprints in the snow on the ice, heading in a straight line for the unfrozen water where the feeder burn issues and prevents the ice forming. We have recorded them fishing here in every month of the year. Sometimes we've seen a bitch with well-grown cubs, sometimes just a single animal, like this one.

We use the term 'habituated' to describe wildlife that ignores human presence, even though it can see people and knows perfectly well that humans can be a serious threat. In this instance the otter must have heard and probably seen the school kids running around in the wood. It would almost certainly have seen me sitting on the sunny deck of the Illicit Still and been well aware that Lottie was pottering about at the water's edge next to the beaver lodge.

'Habituated' is a reasonable enough term, but it does not mean tame. There is nothing tame about this otter – it is truly wild – wild life. What I like to think 'habituated' also means is that because it has visited the loch many times, it has come to expect and accept some level of human presence at a safe distance. But if we were to attempt to approach it, or chase it, or fire a gun, or alarm it in any other way, it would vanish in an instant and probably only return under cover of darkness or not at all. Wildlife of all species run perpetual risk assessments about us humans. They have to, to stay alive.

The mallard duck had seen the otter too and her rasping alarms repeating over and over again echoed over the water. She hurried to lead her brood of nine recently hatched ducklings ashore where they could hide among the rigid stems of bottle sedge, a forest of green rods emerging from deep liquid mud, easy for a duck to navigate through, much harder for a heavy otter. I have seen otters snatch ducklings from underwater, one after another. Easy prey. But we hoped the otter wasn't interested in ducklings today; it was busy fishing for brown trout and there are plenty of those of all sizes in the loch.

It was not always like this. When I first came to Aigas forty-five years ago, otters – along with wildcats and pine martens – were about the most exciting wildlife we were likely to see. Guests were ecstatic if we managed to get them a sighting of one on the river or the Beauly Firth foreshore. People who had been desperate to see a wild otter for years came to us begging to be taken out at first light on a spring or summer morning. No one could have foreseen that in just a few decades they would become commonplace, an everyday sight on rivers and lakes, ignoring people and traffic in many towns and cities throughout the nation.

Today proved to be memorable at several levels – one of those days you never forget. There are moments, only rarely in my experience as a naturalist, when one is dazzled by a particular collision of events so sublime that they catch you off guard and render you wordless with an overwhelming sense of transfiguration.

Lottie had gone about her business. The children's voices had disappeared with them over the rise in the pinewood and the loch fell quiet. I returned to my work. A local man, James Whyte, a member of our little fishing syndicate, is a consummate fly fisherman and visits the loch regularly. James arrived and waved to me as he quietly prepared one of our boats moored beside the Illicit Still: oars, rowlocks, cushion for the seat, his bag of fishing paraphernalia and finally the long fly rod itself, assembled and charged with the appropriate fly for the day. James knows not to talk to me when I'm working. We both smiled and I waved back. The boat gently pushed off and slid away into the middle of the loch, the oars dipping silently as it went.

James had been fishing for some half an hour – time irrelevant to both of us – when I saw him wave. I picked up my binoculars and pulled focus. He was pointing and mouthing 'Otter!' He seemed to be signalling to a patch of reflected sunlight about twenty yards from the boat. It was so bright that my eyes could barely take it in, a reflected sun dazzle so brilliant that I was forced to look away.

Recovering, I looked again slightly to the left of the brightest water. The boat was static, silhouetted against the shimmer on the edge of darker water shaded by the pinewood. The oars hung in their rowlocks and James's rod flicked silently, the cast peeling out in elegant loops before settling gently on the surface. Hovering only a few inches over the dark water were uncountable millions of dancing *Chironomid* gnats, a swarm so numerous that they defied even the crudest estimation. The sunlight reflected from every flicker of wings and bodies gave each insect a mercurial, almost ethereal presence, layering the loch with a mosaic of flashing lights. And there, right in their midst was the otter.

Completely surrounded by the gnats, its glossy head rose to the surface, shining like a wet rock as it eyed first James in his boat and then me, an unmistakable human figure standing watching through binoculars. Pearls of light dripped from its whiskers.

I don't suppose that otter stayed on the surface for more than a few seconds, but long enough for a startling series of images to burn themselves into my brain. When it dived again the hump of its back shone like old silver, then the tail, a gracefully sliding silver signal lifting to near vertical before slipping beneath the dark surface. It continued to fish for another hour before it vanished altogether, and James and I saw it rise and dive again several more times, but never again caught like that, trapped in a blaze of extravagant sunlight and a dazzle of insects, a fleeting moment of unforgettable natural beauty.

Twenty years ago

A June treat. To celebrate the summer solstice and the eighteen-plus hours of daylight, every year I used to take a rowing boat upriver some twelve miles and leave it on the bank of the River Glass at a convenient launch site near the village of Cannich. To wait for the clear weather of an anti-cyclone was essential for a bright dawn and sunrise at about 4.00 a.m. The following morning I would raise whichever of my children was up for it at 3.30 a.m. and drive to the boat. With a packed breakfast and flasks of tea or coffee in a basket prepared by Lucy, we would set off to row gently downstream all sixteen winding miles back to Aigas. It was a rare chance to observe nature long before the rowdy world of humans came barrelling in and disturbed everything.

Rivers are special and we are very lucky to live beside a truly wild river fed by the lochs, feeder burns and tributaries from the glens to the south and west of Aigas. With only a few bumpy rapids it is easily navigable in a small boat all the way down from Cannich to the hydro dam at Aigas. A chance to see roe and red deer close up, perhaps a fox, badger or pine marten exploring the riverbank, always herons, goosander, dippers, mallard, wigeon and teal, and, of course, the possibility of otters. Silently approaching in a boat does not seem to alarm wildlife; perhaps by sitting still we don't look like humans – an image wildlife doesn't find alarming. It is a very enjoyable way to see what's out and about without frightening them away.

I wrote about one such expedition with our youngest daughter, Hermione, when she was nine years old, in my book *Song of the Rolling Earth* (2003). It was special. It was in fact that early morning expedition which spurred me on to write about many more expeditions with Hermione in that book's sequel, *Nature's Child* (2004).

The River Glass has several islands. Glacial valleys with steep walls contain their rivers, preventing floods spreading sideways. Over many centuries the stream has meandered across its floodplain a quarter of a mile wide, veering from one side to the other. When a warm wind blows in from the Atlantic, stripping snow off the mountains, and that meltwater combines with a very wet spring, angry spates have come roaring through, flooding the glen from wall to wall, cutting off whole meanders, creating ox-bow lakes and marshes and isolating strips of bank as random islands trapped between two running prongs of the current.

Ungrazed by sheep or deer, such islands quickly go wild. Willows and alders self-seed and rush to the sunlight; reeds and grasses thrive, creating a tangle of bushes and saplings all thrown together with flood debris. We decided to land on an island to have our picnic breakfast on a grassy spur. It was still only 5.30 a.m. That bright, sunlit morning in 1999 holds an exceptional and particular memory. Hermione was just nine years old:

We find where an otter has habitually left the water and entered it again, sliding down the bank on its tummy. We see prints in the mud – recent too. A stained patch of sick grass and woodrush tells us it urinates there – a liquid signpost to any Lutrine callers. I think it was here this morning.

[. . .] I close my eyes and the sun is still too strong. I turn sideways. Hermione has found some red ants in the sandy soil; she is fending them off with a grass stem. I wonder how ants get off an island.

'Daddy, you're snoring.' I struggle back to the surface. I didn't mean to drop off. I raise myself up onto one elbow, blinking. Hermione is still prodding ants beside me. Past her, over her hip, just there, not eight feet away, the sleep blur

focuses on a face. It is wet and shiny. It sparkles. It is round like an old tomcat's. It has ears like aspen leaves, neat and curved. It has whiskers, stiff and hard, arranged in a downward fan. Beads of water hang from them, catching the sun. Its fur is spiky, as though it has just been rubbed with a towel. Eyes like black pearls peer. It is transfixed. It stands square on, slightly pigeon-toed on short legs; it is broad-fronted like a strong dog. Its body rises behind it to a curved hump and a long sleek tail curls down to the river. From its tip a trickle of water is running back to the stream.

It is looking straight at us.

Thank God we're lying down. We don't look human. 'Don't move,' I growl at Hermione through my teeth so quietly and so sternly that she freezes. 'Otter. Turn very slowly.' Her eyes pass through mine like a cloud crossing a puddle and they keep going, slowly, gently, down the length of my body and out across the river, still panning . . . slowly . . . slowly . . . over her own legs and back into the bank behind her. Her head stops. I know she has connected.

The otter has not moved. It is astonished; it can't quite believe its eyes. Never before has it met a human on this island. It thought it was its own: a place where it can slide in and out without a care. Somewhere to crunch its fish with needle teeth and roll in the spring grass. Perhaps it brings its mate here – perhaps she is the mate and has a holt here, under the alder roots? I shall never know. I know that any second now, he or she is going to turn and slip back into the river with scarcely a ripple. It will re-enter the river by melting. It will vanish in a ripple-thong. It will leave only the mark of its five spread toes in the sand and its liquid image seared into the quick of our singing amygdalae. I know that this is one of those million-to-one chance encounters that gild the lucky. Hermione may not see an otter like this again for years. I hold my breath.

I don't know how long we stared at that motionless otter, Gavin Maxwell's 'friendly daemon . . . [who] . . . put vetoes upon my reason and sent me to look for berries in the proper season.' It cannot have been much more than a minute. We saw its nose rise and its nostrils widen a fraction as it pulled us in, sucked in our shedding molecules and dragged them over its fizzing sensors. Without any hint of panic or alarm it turned and slid into the river. The last we saw was its wet, humped back disappearing below the rippling current and a thin line of bubbles heading away downstream. In a whole minute you can absorb a lot of otter; and an otter can absorb all it is ever likely to want of man. Neither Hermione nor I will ever forget that otter. It will never venture to the island in such happy innocence again.

There was nothing to say. We returned to our boat and set off down the river once more. We had many miles of the long, still pools of the river's maturity to explore.

Living with otters is not about seeing them. They are secretive most of the time and often nocturnal, slipping soundlessly in and out in the darkness. The sightings we do get are in daylight, often brief and utterly unpredictable. But we do know that they visit us all the time. We catch their images on camera traps under the bridge at the loch dam. We see their fishy spraints on habitual stones at the edge of the burn; their footprints appear in muddy paths, and not infrequently we find the evidence of their kills. A trout hauled ashore and partly devoured; frog and toad skins floating in pools; even the remains of our domestic ducks.

Lucy has forty hens, twelve guinea fowl and about a dozen Call Ducks, Khaki Campbells and Indian Runners. They have their own enclosure, pond and secure hen house and we religiously lock them all away at night, every night, to protect them from devastating raids from foxes, pine martens or even badgers – and, very rarely, otters.

Not so long ago we had a favourite duck of inde-terminate domestic lineage (I think you can trace almost all domestic ducks back to mallard). It was pure white. If it had a name I have long since forgotten it. For the purposes of relating its demise I shall call it 'Jemima' – it certainly shared the witless naivety of Jemima Puddle-Duck. One spring evening Jemima had discovered a hole (possibly ripped open by a badger) in the wire netting of their enclosure and she waddled off to explore. She crossed the farm track and discov-ered the burn. Doubtless the attractive tinkle of water over stones was very inviting to a duck. The pools will have contained succulent aquatic insect larvae and other such delights. What Jemima did not know was that most nights the otters left the river and travelled up the burn to fish in the loch.

Normally the otters were not interested in ducks or hens, although well capable of taking them. The domestic birds' association with regular human activity was probably sufficient deterrent to make them keep their distance, even though the burn passed within a few yards of the perimeter fence.

Lucy didn't notice that Jemima was missing that night when she did her rounds, called them all in and locked them away. Jemima was happily puddle-ducking somewhere out of sight in the burn where it runs parallel to our farm track to the loch. Night softly fell. An otter came up from the river, following the burn where a culvert takes it under the main road, on upstream through a dozen pools, through the weedy farm pond until it reached a second culvert beside the track. As it emerged from the culvert, there was Jemima as bright and glaringly visible as a white flag of surrender. No more Jemima.

In the morning I found the evidence. A flurry of white feathers fluttered on the banks of the burn and at the loch I found a fresh otter spraint with yet more feathery content. Of the corpse, the head, beak, feet or carcass, there was no sign, which didn't surprise me because there are so many other guests at a feast like that: buzzards, red kites, pine martens, foxes and badgers, and above all the ubiquitous and extremely observant hoodie crows, are all capable of snatching whatever morsels they can and making off with them. Nowadays Lucy counts her birds in at night.

10 July: 2.30 p.m.

A morning of calm and bright sun. A day of promise, so quick to unravel. My mobile rings. Louis, one of our young ranger team, reports that a guest has sighted an otter on the bank of the loch. He thinks it's badly injured and had blood on its head. 'Is it still there?' I asked. 'I think so,' Louis replied.

We often have otters visiting the loch. There is a healthy stock of wild brown trout in it, but we supplement them with a few farmed rainbow trout, put in each spring for our little fishing syndicate of local boys. It doesn't take long for otters and ospreys to discover easy fishing – sightings greatly enjoyed by our field centre guests.

I went to look. The otter was very visible, curled up on the bank. As I arrived it slipped into the water and dived before I got a proper look at its head. It swam underwater heading for the other side of the loch. Otters are powerful swimmers. As they swim they exhale a stream of bubbles, often all you get to see of them. They are expert at vanishing. I watched the bubble trail through binoculars. It seemed to me to be slower than normal, bubbles breaking closer together than I would have expected as it headed across towards the Illicit Still, where the deck and jetty jut out over the water. I ran quickly round the bank to be there if it came ashore again. Our friend Marion Marshall is a keen wildlife photographer and was on the deck, sitting out in the sun. 'Otter,' I whispered as I arrived, pointing. 'Heading this way.'

The otter made landfall underneath the deck. We could just determine its movements through the gaps between the planks. In our clamour to see it I think we disturbed it and it re-entered the water and swam away underwater. Again we followed the bubbles. I was now certain the bubbles were closer together than usual and that the otter was swimming more slowly than normal. I also noticed that the bubbles were surfacing in clusters, whereas usually an otter stream is of evenly spaced single bubbles. It suggested to me that the animal was struggling with breathing, its exhalation irregular and uneven.

To our surprise, after heading out into the loch for a few yards, the otter turned sharply right and headed back to the bank, a further indication that it might be injured. It came ashore only fifteen yards from us, climbed up a mossy bank and began to rub itself dry. Marion and I were spellbound, thrilled to have such a close encounter with a wild otter – until we saw its head.

'It looks as though someone has hit it over the head with an axe,' Marion whispered to me. And it did. An open wound gaped from just above its left eye, stretching back to beside its left ear. I was watching through binoculars; Marion following through the powerful telescopic lens of her camera. The shutter rapid-fired five, six, seven times. The otter was rubbing its head wound on the moss; I'm sure it knew we were there but didn't seem to care. It was clearly unwell. Marion pulled up the photographs on the camera screen and blew them up. She gasped. We were both speechless with horror.

A story begins to unfold. Back in 1990 the newly created government agency for the natural environment in Scotland, Scottish Natural Heritage (SNH), engaged in a public consultation about reintroducing the Eurasian beaver, *Castor fiber*, to Scotland. The public was 69 per cent in favour. For sound ecological reasons beavers had been successfully reintroduced in a dozen European countries. SNH then recommended a reintroduction trial to the even newer Scottish Parliament. Knowing it would take politicians several years to get round to approving such a trial, at Aigas we were encouraged to commence an enclosed beaver demonstration project at the Aigas loch, to which interested groups and stakeholders such as farmers, foresters and fishermen could come to learn about beaver ecology and see what impact beavers had on their habitat. Before going ahead we had the loch checked out by expert beaver scientists. It proved to be an ideal site, perfect beaver habitat.

In 2006 we imported a pair of Bavarian beavers from an English wildlife park. They quickly settled in and built themselves a massive lodge; over the next few years it accumulated an estimated twenty tons of logs, sticks and mud in a huge oval mound. They bred. By 2008 we had an established beaver colony living wild. Their large enclosure was ring-fenced, containing some fifty acres of mixed woodland, but beavers rarely go more than fifty or sixty metres from their water body, so they probably never knew they were captive. Beavers are very territorial. To prevent them fighting and killing each other we caught up the adolescent kits at a year old, before they became sexually mature, and shipped them out to wildlife parks in England.

When back in the 1990s I first started studying beavers in Norway, I imagined that a beaver lodge had a large central chamber shared by the parent pair and their young. I quickly learned that was far from the case. Lodges have an underwater entrance leading to a small ante-chamber where they groom themselves and store food, but after that the lodge diverges into several narrow tunnels leading to small, well-spread internal chambers where beavers sleep and the kits are born – a survival tactic so that if a predator such as a brown bear or a wolverine digs down into the lodge, it would have difficulty finding the chamber with the kits in it.

Kits are born small but fully furred, well-developed in the uterus. They don't emerge from the lodge until about six weeks old so are vulnerable to predation for all that time. It is well documented that unlike powerful diggers such as bears and wolverines, otters can sneak into a lodge by the underwater entrance and sniff out the kits. Fact: otters regularly kill and eat baby beavers.

When people ask me 'How big is a mature beaver?', I reply: 'Think fat spaniel.' A beaver is a formidable rodent sometimes weighing up to 35kg. It also has fiercely sharp, chisel-shaped incisor teeth and very powerful jaws capable of felling large hardwood trees. You can see where this tale is heading. Our otter had made a terrible mistake.

Marion and I examined the photographs carefully. There were unmistakable beaver incisor teeth marks on the top of its head, and the open wound, right through the fur and the flesh, right through the bone of the skull and the meningeal membrane, precisely matched the size of bite a beaver takes out of a birch tree. That unhappy otter's brain was clearly exposed in an oval the size of a ginger biscuit. The rest of the story is speculation, but has a high probability score.

That wound is extremely unlikely to have been inflicted in any open space. I have witnessed many beaver–otter confrontations on the bank, and they come to nothing. The otter is too quick and agile to be caught by a beaver and anyway the beaver is not aggressive, preferring to slip back into the water and disappear. Beaver and otter scientific literature is full of such encounters. No otter in its right mind would pick a fight with a large adult beaver. But it is also well documented that otters do enter beaver lodges when the adult beavers are absent. We believe that is what happened.

The big matriarch beaver – we have never weighed her, but she is big – entered the lodge when the otter was inside and they met in a tight tunnel, face to face. The beaver had filled the tunnel, blocking the otter's only exit. In a panic to escape it probably tried to push past the beaver, or even underneath it, giving the beaver the chance to take a huge bite out of its head.

That otter was mortally wounded. No mammal can survive for long with its brain exposed to infection. Little wonder it was having breathing problems and was rubbing its head on the moss. I am sure it died quickly, probably somewhere out of sight, curled into one of the many mossy hollows around the loch. It would have been good to be able to do an autopsy, but despite extensive searches we have never found the corpse.

June

The summer solstice has slid past us without anyone really noticing. Long daylight means our guests can see pine martens, beavers and badgers out and about in broad daylight, even in sunlight, well into late evening. This week the field centre is buzzing with three simultaneous courses: one on Aigas Wildlife; a High-level Hikes programme exploring the splendid Highland mountains; and one on Wild Flowers. I offered to help out with general wildlife.

Visitors to Aigas at this time of year hope to see all the highlights for which the uplands are famous: golden eagles, peregrine falcons, red and roe deer, pine martens, beavers, otters and ospreys, and masses of other birds. On a birding week we regularly record well over a hundred different species. People get very excited about seeing a local speciality: Slavonian grebes are always a favourite, with their outrageous marmalade eyebrows and intricate mating dances.

Dawn is absurdly early. There is a marked reluctance among our young ranger team to lead early morning forays to the coast of the Beauly Firth, fifteen miles away, so I offer to lead it for them. The guests are keen, up for anything that will deliver the goods. Even when I tell them we need to be on the road by 3.30 a.m., they are still nodding, if not quite so enthusiastically. Dawns are somehow written into my DNA. I love them. I am more awake at dawn, more alive and alert, whatever time it arrives, in any season, than at any other time of the day.

I lie and watch the light rolling in from the east as silent as a mist. The booming hall clock strikes just twice. Our bedroom window faces south, so the steely dawn rays creep across the lawns while the room stays in darkness. It will be another hour before it's properly light indoors. Lucy is still deeply asleep. I ease out of bed as a seal slides off a rock on a rising tide, grab my clothes and get dressed in the corridor so as not to wake her. The two dogs, Buster the black Lab and Tuck, the Jack Russell, greet me in the kitchen with frantically wagging tails. Buster throws in a few on-the-spot pirouettes just to make the point. They also love dawns. I nibble an oatcake while the kettle boils, then tea.

Outside is cool and still. The sun is still not quite up yet but I can see its incipient flare highlighting the serrated forest fringe on the hill between Aigas and the Black Isle, where we will be heading later on. On mornings like this Wordsworth's great Westminster Bridge poem floats into mind:

> Earth has not anything to show more fair:
> Dull would he be of soul who could pass by
> A sight so touching in its majesty

The dogs rush ahead, pissing merrily on every other thistle and tuft of grass. We head through the gardens for the fields to see Queenie, my huge Clydesdale mare. She is down, legs tucked under her, eyeing our approach through a sea of daffodil-yellow, rhizomatous creeping buttercups, damp with dew. The other horses are up and quietly grazing a few yards away. The dogs are still snuffling around on a mission of their own. Queenie ignores them. Two ravens pass high overhead, rowing on flat effortless wings, cronking disapprovingly to each other.

In the bottom corner of the field, out of sight of Queenie and the dogs, some two hundred yards away, I spot the trim silhouette of a fox. Through binoculars I watch the easy, tripping lope as it floats along the edge of the meadow, pausing now and then to sniff at something in the grass, a fleeting effigy of the dawn. I wonder if it was the fox the ravens were cronking about. Ravens don't miss much. Seconds later it is gone, vanished into the thick hawthorn hedge that shields the fields from the main road. I glance at my watch. Nearly three o'clock. Time to head back.

My seven guests arrive at the minibus a little bleary-eyed, but dead on time. We climb into the trusty old Volkswagen and slide silently down the drive and out onto the main road. It takes a little over half an hour to reach the foreshore. My passengers are mute. A middle-aged lady, a schoolteacher from Musselburgh called Gail, has fallen asleep, almost immediately crashed out. I guess she doesn't often do dawns. One or two roe deer flounce away from us, skipping daintily back into the woods as we head out of the glen.

When we arrive at the firth we glide gently along the single-track road beside the beach for a further half hour, stopping every few yards to search the shore. The sun is up and lifting fast. From the east its rough, burning light is spreading across the calm surface in waves of mercurial brightness. Mud banks shine like wet metal. The air rings with the insistent sadness of wader calls. Close to us a few curlews are picking through a dark tideline of drying weed. Out on the mud flats, between lagoons of shining water, a medley of wading birds is nonchalantly idling: oystercatchers, curlews, bar-tailed godwit and one or two greenshank. We stop and get the telescope out to scan. Gail has woken up but looks very bleary. She struggles to focus her binoculars. Well out in the firth, cormorants are hauled out on a mud bank in a long line, hanging their wings to dry like heraldic griffins. There are also some tiny waders far out on the tide's edge, dunlin perhaps, but too far away to see, and beyond is the sea haze of distant hills. We move on.

No wind. The tide creeps towards us in tiny, lapping ripples catching the sunlight in thin, parallel lines of silver string. The air is fresh and salty, and the piping calls of oystercatchers mingle with crying black-headed gulls and an occasional curlew's bubbling song. The guests are chatting amiably; they were pleased to see a greenshank through the scope. It was static, perched on one leg, the other one tucked up. 'It's only got one leg,' Gail had muttered. 'Don't be daft,' said Ben with a friendly chuckle. He's a retired railway engineer and a keen birder. "E's restin'. That's 'ow they kip.' They clambered back into the van and we drove slowly on down the firth.

A little while later, out of the corner of my eye, far out on the water and well ahead of us, I see a black, bobbing head in slightly more ruffled wavelets out in the open firth. Too far to be sure what it is. A harbour seal perhaps? I decide to say nothing; we're heading that way.

I know they would like to see a harbour seal, but I want it to be an otter. I have a gut feeling it could be. It is small, just a dot, barely visible between the gentle undulations of the flooding tide. I drive very slowly. It has vanished but has to be there somewhere. There it is again, slightly closer now. I swing into a passing place and scan through bins. It is an otter but still a long way out. As it dives again I see the unmistakable lift and near-vertical slide of the disappearing tail. An otter fishing. I know that is what my people would really like to see. A sharp-eyed pensioner called Mike speaks from the back. 'There might be a seal out there,' he says in a voice not quite sure of itself, pointing. 'Could be,' I reply. 'Let's take a look.' I don't want to get their hopes up.

An unofficial gravelly layby up ahead leads directly onto the shore. It looks as though it's where some local angler occasionally launches his boat. It is as close to the otter as I think we're going to be able to get. 'Out nice and quietly, please.' The door slides back with a low rumble. We line up against the side of the minibus to break up our shape. It's good to have something to lean against when using binoculars and it hides our human form, acting as a bit of a hide. I mount the scope on its tripod.

'It's not a seal, it's an otter.' I whisper the words unnecessarily to encourage them to be as quiet as they can. A rustle of excitement issues like a frisson passing through. Jenny from Helensburgh says 'Where? Where?' Mike guides her to the ring of water where it has just dived.

While it's down I mount the scope again. 'Will it come closer?' Gail asks. 'Hope so,' I mutter. The otter stays down for a long time. It feels much longer than it really is, perhaps a minute and a half. Then up and straight down again before I could get the scope onto it. Something down there is keeping it busy. There follows a succession of frustratingly rapid dives, so rapid that one or two of the group fail to get their binoculars onto it in time. 'Not again,' Gail sighs.

We watch the otter fishing apparently unsuccessfully for another twenty minutes. A posse of common gulls passes over and spirals away to the clouds. Very slowly our otter is working its way closer to shore. 'Ooh! Look!' Another man, George, who is a landscape gardener from the Lake District, is pointing off to the left, much closer. I can sense the rising excitement. It's a second otter, also diving.

'You all watch this closer one, and I'll keep an eye on the first one.' They seem happy to comply. I give up with the scope and just watch through bins. Up. Down. Up again, down again. Both otters slowly working their way towards us. I have a hunch that they might come ashore. Excited low-level chatter from the group. Minutes tick by. Then it all changes.

My otter surfaces with a fish. I can't make out what it is, but it's a good catch, perhaps a grey mullet. I can see the fish squirming in the otter's jaws, tail frantically flipping. Or it could even be a small grilse salmon. It's a decent size, firmly gripped amidships in the otter's impressive, sharply pointed fangs, especially evolved for grabbing fish. 'Mine's got a fish. It's heading this way.' All attention turns to my otter.

Otters struggle to eat fish at sea. I have watched them diving for crabs and munching them on the surface before diving again. But the bigger the fish, the more it is likely to have to come ashore to eat it. By chance we are positioned just right. Both otters are heading straight at us. It looks as though mine is going to come ashore right in front of us. The mullet – if it is a mullet – is protruding from both sides of its jaws. We watch in silence until Gail finally gets her binoculars focused on it and exclaims 'Oh my God! It's coming this way.'

The second otter lands first and starts to rub itself dry on hummocks of bladder wrack. Every few seconds it sits up and peers penetratingly at the otter with the fish. It is only about thirty yards in front of us and has no idea we are there. The first otter is swimming on the surface, heading in. It's obvious these two otters know each other, and I tell the group I think the one with the fish is a bitch and the second one last year's well-grown cub. It's only a hunch but seems likely. Cubs, especially female cubs, can stick around with their mothers for more than a year.

In a slow, looping flight, a heron comes low along the shore about fifteen feet above the tidal edge. I saw it coming a long way off in steady, measured wingbeats with its neck folded tidily under its long sharp bill, feet straight out behind. 'Watch the heron,' I mutter at the group. The big bird glides in as if to land and then sees the otter on shore. A sharp, grating curse breaks from its bill and in a flurry of undignified wings it swerves away and out over the water. I got the distinct feeling that heron did not like otters.

Although I've seen wild otters hundreds of times, I still feel a rising sense of thrill, thrill for my guests as much as for myself. Wild otters grab you; they demand your attention and don't let go. They have pizzazz; they ooze charisma. You can't ignore an otter. We certainly can't take our eyes off these two. The group falls silent and completely still. The otter with the fish has landed and is heading up the beach. 'Keep really still with your cameras,' I urge. We can now see the size difference between the two animals. My hunch may be right.

A hoodie crow wings in and pitches on a rock a safe distance away, the sunlight gleaming on its strong black bill and mantle in dove grey. Hoodies miss nothing. They are villains and arch scavengers, the vultures of the north. This one has seen the fish, hoping for pickings. If we all perish in a nuclear holocaust, when the dust settles I'm sure there will be hoodies picking over our charred corpses. It stands and watches, biding its time.

Every once in a while, fortune favours the diligent; those who are patient and make the extra effort do, every now and again, reap real rewards. It was a big ask to get them out of bed at three-thirty. I feel a glow of satisfaction that it's paying off. They will remember this for years. The bitch otter scampers up the beach with the mullet, the adolescent cub in hot pursuit. (I'm sure it was a grey mullet, they are common in the firth and we see ospreys carrying them in to their Aigas river nest.) At first she seems reluctant to share her catch and turns away, circling so that her cub can't get at it, while coming ever closer to us. We can hear the cub uttering soft, squeaky whistlings as its mother approaches. It seems to be begging for a share. The bitch is now only the toss of a stone away from us, perhaps ten or twelve yards, no more. 'Keep very still,' I whisper. Camera shutters click and whir.

We hear the crunch as the bitch bites firmly into the fish's spine just behind the gills. The cub is close and whickering softly. The bitch takes a huge bite, shakes its head and a chunk comes away in its jaws. It turns away to munch it. The cub quickly grabs the rest of the fish. It has its back to us, but we can hear the flesh ripping and the wet mouthing of jaws. That cub is hungry. The mother turns back and grabs the fish by the tail. A brief tug of strength follows as the fish is torn apart. Both otters now have a decent chunk of mullet. The head has separated and lies on the weed beside them. Black eyes gleaming, the hoodie hops closer.

At that moment Mike the pensioner does his best to stifle a sneeze – only partially successful. Both otters stop eating and look up. The bitch stands up, vertically on its back feet like a weasel, and peers in our direction, uncertain, probing. Its whiskers flare like a fan. It had no idea we were there and now can't quite make out what it heard. We freeze. For a moment I think we have blown it, but in a flash of black bill and feathers the hoodie darts in and grabs the fish head. It wings away to a distant rock with its prize. It has saved us.

Both otters' attention has turned back to what is left of the mullet.

We watched the otters demolish that fish for eleven and a half minutes. For almost a quarter of an hour one of nature's wildest creatures wowed us with a rare and privileged show of ottery feeding behaviour, so very close, closer than we could ever have hoped or prayed for.

They both rolled exuberantly in the weed for several minutes. Rubbing their backs and sides, pushing their heads and necks along first on one cheek, then the other, squirming into and under the banks of seaweed and emerging again with wrack and tangle on their heads like crazy coiffures. They scratched and rolled some more. Then they came together and did some mutual grooming. The cub curled up in the sunshine beside a rock and looked as though it was going to sleep. Its mother sat beside it and stared out to sea.

That rolling was functional, I am sure. Otters have to dry and de-salt themselves continually to keep their fine, silky fur waterproof and in good condition. But to us silent watchers the rolling spoke of much, much more. It told of bodily contentment after a successful feed, of the bond of maternal and filial tenderness, of the absolute joy of two wild, carefree creatures oblivious to the presence of man, unharassed, unthreatened, at peace and at one with their natural habitat.

After a few more minutes they slid quietly back into the firth. The last we saw of them were two shining black heads bobbing away through bright sunlit ripples. In silence we watched them round the curve in the shoreline until they disappeared from view. It was five past six.

6

Death

Two deaths. The extraordinary freak death of an otter by a beaver bite, and of a young sow badger killed on the road. The first an accident of fate, the other the consequence of man's crushing dominion over not just badgers, but all wildlife. I am not shocked by either of those deaths. I am saddened that a young badger in the prime of life was so summarily extinguished by a vehicle, and sorry that the healthy young otter probably had to suffer a slow and perhaps very painful end. Of course I am. You cannot work with wildlife for half a century and not care about individual animals. The badger's death was avoidable and the sadder for it. The otter's death was nothing directly to do with me, except that it was I who introduced beavers into the loch. But since beavers and otters have been cohabiting lakes and rivers for many millions of years, I can hardly accept responsibility for one accident of fate. But of course, biologically interesting though it is, I wish it hadn't happened. A sad end to a beautiful and exquisitely evolved otter.

Steady, now. Humans are an intensely emotional species and most of us are regularly stirred by so many of the twists and turns that, for better or for worse, shape individual lives, our own as well as of others. But nature couldn't care less. It is sometimes important to consider that. And anyway, it is not the end of the story; it is merely the end of the episode.

Evolution doesn't care a jot about you or me or the otter or the badger, or how they died. Nor will Nature care a hoot about how or when you or I die. Nature is bigger and way more heartless than all of us and all our emotions put together. Pain and cruelty just don't feature in Nature's grand plan and the old cliché remains wholly true – 'Nature is red in tooth and claw.'

We humans value the individual supremely and it pleases us so to do. We know pain and suffer it both personally and vicariously. We are all individuals with individuals' keenly honed emotions. But in nature you and I simply don't count. The individual is entirely expendable, a universal truth we find deeply uncomfortable and prefer to ignore. It is one of the main things that differentiate us from most of the rest of nature. Nature is lucky – it is entirely pain- and care-free.

Emotions hold us together. We have evolved them for good reason. They are the electricity of each individual's fizzing psyche. 'In emotion resides the soul,' said Aeschylus thoughtfully, five hundred years before the birth of Christ. As the father of Greek tragedy, he knew a thing or two about emotions. A mother cares 'emotionally' for her child. Just as we can be sure an otter bitch cares 'emotionally' for her cubs. Ethology, the science of animal behaviour, has come a long way in my lifetime. At long last, we have come to realise and accept that most higher animals have similar feelings to our own.

Whales and dolphins, elephants, chimpanzees and, yes, otters, express grief and mourn their dead. So, I'm sure, do a host of other creatures to a greater or lesser degree. Intensely social animals such as gorillas and chimps openly display their emotions and rely on them to function as a group. It is one of the great paradoxes of evolution of the natural world that Nature cares nothing for the individual but has invested such powerful emotions in so many of us. Perhaps that is our true role in life – to do Nature's caring for it.

We need to chill. There are times when emotions block our logical sensibilities. The badger in its three-foot grave and the otter, probably a corpse curled up in its bankside hollow, are both integral elements within Nature's grand plan – that of returning nutrients to the soil, which is where they came from in the first place.

If, now that several years have passed, I were to dig up the badger, all I would find would be a skull, some bones and hair and its teeth – the last bodily components to break down. The whole of the rest of the badger would have fed many millions of other organisms in the process of decomposition. Living soil is composed of billions of micro-organisms. (Recent research has revealed up to 200,000 enchytraeids, 80,000 springtails and 30,000 mites per square metre of soil, as well as uncountable millions of thread-like nematode worms.) Worms and beetles are the vanguard; they kickstart the process, but in no time the soil surrounding the corpse would be a heaving bouillabaisse of microscopic animals feeding, digesting and excreting the badger's flesh, brains and juices until they had *apparently* disappeared.

Don't be fooled. Nature is much smarter than that. Nothing disappears. It just gets recycled. Those nutrients and complex organic chemical compounds may never again directly reassemble themselves as a badger, but you can be certain that sooner or later some of the worms and bugs the badger's corpse fed and fattened will be eaten in one form or another by other badgers, possibly even badgers related to the dead sow, thereby completing the most fundamental cycle of all life.

Even as I write, only a week from seeing the doomed otter with Marion, sexton beetles will have sniffed it out and have homed in to lay their eggs in the corpse to feed their larvae. If the otter's chosen tomb happens to be open to the air, blow flies will have rushed to the party and laid their hundreds of eggs on that exposed brain, on the glazed eyes, ears, nose and lips – any exposed soft tissue. In summer, fly eggs hatch in a matter of a few hours; the otter's skull will likely now be a squirming frenzy of hungry, greedy maggots all busy consuming and digesting flesh and brain, thereby diligently executing Nature's utterly heartless, reductionist plan. I'm sure that, just as with the badger, sooner or later some of those flies will be snatched by trout in the loch, some of which, sooner or later, in the fullness of Nature's grand plan, will be caught and eaten by a passing otter. Life and death are both pawns of chance in the game of survival. It is well to remember that.

★　　★　　★

A naturalist at large does not just observe and identify wildlife, but also must endeavour to learn about its complicated habitat requirements – its oh-so-particular behaviour and growing, breeding or nesting needs. After nearly fifty years of trying my best to find answers, all I find are questions. Questions spawning ever more questions. That's what makes natural history so fiendishly fascinating. Even after decades of watching and wondering, after many thousands of hours in the field, after days of sitting still, hunkering down in Zen-like mindlessness, the questions keep coming. They never stop. They are infinite.

Why do certain plants only grow in specific habitats? In June every year a heath-fragrant orchid, *Gymnadenia borealis*, appears above the loch, just so. Just in that place, where I placed a tiny wooden marker years ago, does it choose to erect its perfect stem and its thirty perfectly exquisite little lilac-pink florets packed together in a neat little cone-shaped inflorescence, each one precise in every detail, and all emitting a heady bout of perfectly wondrous evening scent of crushed oranges with a hint of vanilla and chamomile. Always in the same slightly marshy, grassy spot, never three feet to the north, south, east or west, and nowhere else for miles. Why? What is going on here? Doesn't it want to set seed, spread and proliferate like just about every other plant I can think of? Who is ordaining its perfectly precise little habitat? Beats me.

Why are some butterflies and moths so intractably wedded to certain food plants for their caterpillars? I remember discovering and being amazed that the orange-tip butterfly only lays one tiny orange egg, smaller than a pinhead, on each stem of the common meadow wildflower lady's smock (often called cuckooflower). Why not more? Is it that the caterpillar actually needs the entire flower to itself? Or is it because it might well be munched by a cow and better to lose one egg than many? Could the female butterfly really know not to risk all her eggs in one cow? Pass.

Why do our pine martens need to gorge on rowan berries every autumn? Is their very catholic omnivorous diet really so short of vitamin C that they have to concentrate exclusively on those acidically bitter berries for as long as they are ripe? (Don't try them, they are disgusting and it will take hours to shed the desiccating bitterness from your mouth.) Why are they so unpalatable to us and apparently so attractive to martens? And what exactly is going on when they choose to position their messy rowan-berry faeces on the very top of fence posts and prominent boulders?

Why are our utterly captivating crested tits, *Lophophanes cristatus*, so completely wedded to pinewood? And ancient pinewood at that. Are there such specific bugs in pinewood that they can't do without? If so, what are they? And why won't the fussy tits or their bugs venture into beautiful adjacent birchwood? What is stopping them?

How do the otters half a mile away in the River Beauly know so very quickly that we release fat and juicy farmed rainbow trout into the loch every spring for our little fishing syndicate of local boys? Is there a fishy scent invitation that travels down the outflow burn, or is it just coincidence and curiosity, hardwired ottery opportunism, that makes them arrive within a day and check out the loch, just in case?

Why are badgers so territorial and so dramatically brutal to each other when they clash or meet, or even when they mate? We almost never see badgers without scarring to their faces, ears, necks and rumps. Is life really so tough that they have to beat each other up? Aren't there enough challenges to being a badger without resorting to mindless domestic violence towards their own kind?

And perhaps the biggest and most pressing question of all, for all naturalists and students of wild nature, is why we humans are so blinkered. Why, after centuries of losses of our precious wildlife, have we not learned to live with the other creatures of the world? Why have we not understood that we need them, not just for our own sanity, but because they offer so many benefits to our own species?

I vividly recall when I was a young student of ecology, the late, great, Sir Frank Fraser Darling telling me that the Latin verb *conservo*, from which we get our word 'conservation', actually means 'I maintain'. Well, we certainly missed that trick! We have an appalling record of maintaining anything in nature. Nature conservation has been one step forward and two steps back throughout its entire history – no hope of slowing down human greed and consumption, even when nature's benefits are plain to see. We know full well how vital habitat is to all wildlife, but with the shining exception of a few nature reserves sprinkled across the nation, the only habitat we have really focused on is our own.

* * *

'To the dull mind all nature is leaden. To the illuminated mind the whole world burns and sparkles with light.'

20 May 1831 (journal),
Ralph Waldo Emerson (1803–1882)

'Nature never did betray
The heart that loved her'

'Tintern Abbey',
William Wordsworth (1770–1850)

Tailpiece

Every morning I awaken in hope. I stand at that open window and stare. You ask 'What are you hoping to see?' I don't know. In truth, I really don't know and I don't think I have ever known. Nor have I ever given it much thought because sooner or later some fragment, some flash or spark of the natural world will come rippling through to grab me and haul me off. That is the way of the wild world. It never sleeps and it always keeps you guessing. Perhaps that's it. Perhaps it is just that brazen unpredictability, the sheer brilliant creativity of nature that gilds the living moment. You do not have to tell me how lucky I am to live and work with wildlife.

My language is that of reportage: of nature's open and honest narrative, brave, bold and startling – never dull, never complacent, never corrupt. It is what I see and feel in the day-to-day existence we glibly pass off as life and work. To me life is a book, chapter after chapter of unending mystery and drama. Every day it falls open at random; I never know what page I'm on. I think I've come to the end of a passage and then it fires itself up all over again. It is a book of yearning and longing, of seeing and believing; it is also a book of hope.

When I was just seven, an old countryman told me that a fox regularly holed up in the hollow of a big oak stump at the end of the garden. I went to look. There was nothing there. A few days later I was passing and thought I would look again. I thrust my head into the void, almost completely filling the hole. The fox shot out. Its nose, teeth, ears and neck, its shoulders and flanks swept past my face, shoving me roughly aside. Its pungent animal scent engulfed my nose, my head, my lungs. It was my first real encounter with a wild animal, up close – far too close – and personal, my first real brush with wildness. I have never forgotten that heart-thumping moment and it carried with it a rush of excitement that has never dimmed.

*   *   *

I would love to have known that seminal trio of nature thinkers, the nineteenth-century Transcendentalists: the poet Walt Whitman (1819–1892), the essayist and philosopher Ralph Waldo Emerson (1803–1882), and their naturalist-philosopher protégé Henry David Thoreau (1817–1862), author of *Walden*, an experiment on 'simple living' in the (relatively tame) wilds of Concord, Massachusetts. But they all had the huge advantage of experiencing a vast continent of unsullied wilderness. When, led by an Indian guide, Thoreau made a failed attempt to climb Mount Katahdin, Maine's highest mountain, he got a shock and a fright at his first experience of real wildness. He returned home a changed man.

> This was that Earth of which we have heard, made out of Chaos and Old Night. Here was no man's garden, but the unhandseled globe. It was not lawn, nor pasture, nor mead, nor woodland, nor lea, nor arable, nor waste land [. . .] Man was not to be associated with it. It was Matter, vast, terrific [. . .] rocks, trees, wind on our cheeks, the *solid* earth! the *actual* world.

It is remarkable that Thoreau's more than three million words of philosophical writings on human interaction with wildness remain in print more than 160 years after his death. Heralded as one of the founding fathers of the modern environmental movement, it is sad that we have so studiously ignored his wisdom.

Emerson's dream haunts me: 'I dreamed that I floated at will in the great Ether, and I saw this world floating . . .'. Little wonder it dazzled him and spun him into mysticism. Now that we really can see our world floating it isn't so much dazzled we need to be, but ashamed. What we have done and continue to do to our world is more than shameful, it is a spectacular denial of the wisdom *Homo sapiens* is supposed to have evolved.

The Transcendentalists' contemporary, that other American luminary, the poet Emily Dickinson (1830–1886) from Amherst, Massachusetts, may not have intended it, but her lines are startlingly apt for today's predicament.

> I held a jewel in my fingers
> And went to sleep
> The day was warm, and winds were prosy
> I said, 'Twill keep'
> I woke and chid my honest fingers,
> The Gem was gone
> And now an Amethyst remembrance
> Is all I own.

Our work at Aigas is to do our best to unravel some of the mysteries of nature and to share them with our guests as they really are in the wild. Our strapline is *Sharing the Wonders of the Wild Highlands*. It means sharing the good with the bad, the uplifting with the shocking, exposing the thoughtless hand of man's ruthless and careless domination of nature's forgiving and healing beneficence, as well as celebrating her ceaseless and omnipresent creativity. We want our guests of all ages to live in the moment, to see, feel and absorb a few fragments of what we have left. We want them to take away gems of their own.

14 September

It is autumn. Only a week away from the equinox, daylight is shortening by six minutes a day. It's getting dark now when we put the ducks and chickens away at 7.30 p.m. Lucy worries about carnivorous wildlife out on the prowl – with good reason. She remembers Jemima Puddle-Duck only too well. Yesterday evening a badger tipped over the kitchen bucket of scraps put outside by our Polish cook, Genia, to be taken to the chickens. The badger guzzled most of its contents. As Hermione walked past it to her car to go home at nightfall, it took no notice. A wily old badger wouldn't think twice about grabbing an unwary hen.

Autumn. John Keats's much-fêted 'Season of mists and mellow fruitfulness'. No frosts yet, but the bracken seems to know what's coming. It's browning off and collapsing in a tangled heap. The heather flowers are past; just ginger shells on stems that were vibrant with purple uniformity only a week or two ago. Almost all the wildflowers are over; the exceptions are the blue-mauve button heads of devil's-bit scabious among stars of bright yellow hawkweed, the last splashes of tech-nicolour among fading grasses. The trees have yet to turn, and we are waiting for the first stag's roar of the rutting season – any day now.

I had been working in the Illicit Still all day, a day of late summer sun and warm breezes. Needing a leg stretch, at 3.30 p.m. I called the dogs and set off on the trail around the loch. Buster the Labrador ran ahead, tail gyrating with happiness, and Tuck, the Jack Russell, ageing now, toddled happily along at my feet. I'm glad he doesn't know he's fourteen.

This is the trail our guests enjoy most. It hugs the loch shore, passing through pine forest where crested tits and crossbills grab the limelight, onto lovingly built boardwalk across the marsh – the only way we can protect it and let folk experience the wetland at the same time – and on into birch woodland, which is spreading up the hill like a quilt caressing the land. The trail lifts onto the heather moor and drops down again to the huge beaver lodge where the otter is most likely to have received its mortal wound.

The loch is dark and tranquil in the shade of the pinewood, ruffled with shinkly sunlight over the rest. I never tire of its shifting lights. The dinner-plate water lily leaves have browned and sunk out of sight and the lanceolate spears of *Potamogeton* will soon follow suit. Even without a frost, long fronds of birch are yellow-gold among the green. Autumn's wild pallet is every-where I look. Every few yards a flush of fungus has burst through: *Russula*'s lilac discs, glossy brown 'penny bun' boletes, fly agaric's handsome scarlet-and-white spotted caps spelling 'poison' loud and clear, and the gleaming waxy ochre of chanterelles – delicious edible gold – among a host of others whose names I have long forgotten.

There is nothing mellow about the fruitfulness of the rowan trees. They are bursting with clusters of crimson berries, some so heavy that they have bowed the branch ends almost to the ground. A pine marten has been here too. They climb out as far as the branch will support them, bite through the stem heavy with berries and then jump down to the ground and gorge. They gorge and gorge for as long as the berries last. The result is scattered at my feet, berries cast in all directions – a seasonal berry-fest. It happens every autumn and I'm not sure anybody really knows why. Over the next few weeks mixed flocks of chattering Scandinavian thrushes – fieldfares, redwings, mistle and song thrushes – will flood through, stripping the berries the pine martens can't reach and then ejecting the fertile seeds as their migration moves on; rowan trees are sprinkled right across the landscape. The birds don't know what splendid agents of seed dispersal they are.

The trail passes through our first restoration ecology project, established more than forty-five years ago. When we arrived here the moorland behind the loch was as bald as a plate. It had been deforested hundreds of years ago to create much-needed grazing for Highland folk. Without leaf-fall to nourish the ground and build soil, all fertility had slowly drained away. Centuries of overgrazing had removed all the nutritious grasses, leaving only the unpalatable acid-loving others. Successive generations of cattle, then sheep and finally deer had prevented any natural regeneration of the trees that are such vital agents of natural restoration.

When we arrived in 1976 we removed the grazing altogether, controlled the deer and let nature back in. The result forty-five years later is broad acres of all-singing, all-dancing native woodland and wetland of eight or nine native tree species, sixteen species of nesting spring birds, a host of wild flowers, countless invertebrates, all the mammals: roe and red deer, foxes, pine martens, badgers, red squirrels, weasels, wood mice, beavers and water voles, field and bank voles, shrews, including water shrews, three species of bats, visiting otters and an invisible network of mycorrhizal fungi and other fungi, ferns, lichens, mosses and liver-worts, all existing together in a dynamic woodland ecosystem. Uncountable hordes of invertebrates have motored in and made it their home.

I love it and bow my head to it every time I look across the loch. It is what we teach at Aigas – restoration ecology – giving back to nature and allowing it to heal itself. At the Illicit Still a few Highland darter dragonflies with blood-red bodies and rigid wings are basking, soaking up late summer sun on the rails around the deck. They are the last dragonflies of the year.

When I get back to the house I learn that a savvy little weasel has sussed that wood mice regularly come in to the feeding station at the red squirrel hide. It lies in wait in the drystone wall, and watches. A weasel ambush is fizzing in its sharp little head. Then it nips out to grab a passing mouse; this afternoon, Richard, a guest in the squirrel hide with camera poised, enjoyed a moment of rare fulfilment. He grabbed the chance to capture its portrait.

The badgers are hurrying now. They emerge at dusk and spend the long dark hours foraging to lay down brown fat for the winter. When the snows come and frost falls on moonlit nights, badgers can't feed for days on end. They have to rely on that fat for survival. Little wonder the badger at the kitchen door had conquered his fear of humans and was feasting on the rich pickings from the hen bucket.

We haven't seen an otter at the loch for some weeks now, but they still appear regularly on the river and the foreshore of the Beauly Firth. I know very well that when we restock with trout next spring, they will be back, as they have been every year that I have lived in this bountiful place. So will the ospreys, and the sparrowhawks will raise their squeaky young in the pinewood. Roe does will slip their twin fawns hidden in among the spiky juniper tangle at the back of the loch and the sow badger will bring her boisterous cubs to the hides to feast on peanuts. I hope the weasels will nest again in the old wall, and I will struggle all over again to find the martens' dens.

Tomorrow we have a large class of primary school children in for the day; more than twenty, I'm told.

I look forward to hearing their rowdy laughter as they progress round to the marsh for pond-dipping. They are our future. I want them to be able to enjoy wildlife as I have in my time. We have so much work to do.

John Lister-Kaye
House of Aigas, 2021

Acknowledgements and thanks

Naturalists often choose to work alone, especially when challenging stealth and patience are required, but living through a life-long career at a field studies centre such as Aigas has meant that I am constantly surrounded by keen young rangers and staff, themselves highly competent field naturalists, with whom I can debate and test my observations and encounters.

Over the decades there have been too many to mention, but in particular I would like to thank several of those whose input was particularly valuable to me for this book: Ben Jones, Taylor Davies, Paige Petts, Tom Chambers, Lottie Budd-Thiemann, Calum Urquhart, Charlotte Robertson, Josephine Tod, Eliane Bornoff, Dougie Brunton, Emilie Shuttlewood, Emily Richens, Susie Kerr, Louis Schopp, Dora Hamilton, Lucy Smith, Mathew Broadbent, Michelle Branson, Richard Thompson, Chloe Denerley, Rachel Buckley, Milo Mole, Jonathan Willet and Elizabeth Hare.

My old writer friends Mark Cocker, Stephen Moss, Brian Jackman, Polly Pullar and Jay Griffiths are a constant inspiration and encouragement to me to keep writing. Laurie Campbell and Liz Holden shore me up every year with their dazzling natural history knowledge and skills. I owe a huge debt of gratitude to them all.

I cannot omit the wildlife. We live with pine martens, badgers and otters every day of our lives, and slightly less frequently with weasels. Our field centre guests enjoy hide visits to see wildlife up close and personal every week. I don't think Aigas could exist without them all tolerating us and our often-intrusive activities. At least they seem to know they are safe and welcome here.

Lastly, and unquestionably, my wife Lucy and our daughter Hermione are constant props to my fragile literary ego. They tolerate my moods and irritating habits – Lucy describes herself as a 'literary widow' when I'm writing – and they both regularly overlook my inconsistencies. I could not be a writer without them.